Teaching the History of Science

Teaching the History of Science

Edited by

Michael Shortland and Andrew Warwick

The British Society for the History of Science
Basil Blackwell

First published 1989

Basil Blackwell Ltd
108 Cowley Road, Oxford, OX4 1JF, UK

Basil Blackwell Inc.
432 Park Avenue South, Suite 1503
New York, NY 10016, USA

British Library Cataloguing in Publication Data
A CIP catalogue record for this book is available from the British Library.

Library of Congress Cataloging in Publication Data
Teaching the history of science / edited by Michael Shortland and Andrew
   Warwick.
      p. cm.
   Bibliography: p.
   Includes index.
   ISBN 0–631–16977–6 — ISBN 0–631–16978–4
(pbk.)
   1. Science—History—Study and teaching. I. Shortland, Michael.
II. Warwick, Andrew.
Q126.9.T4   1989
509—dc19

89–33
CIP

Typeset in 10.5 on 12 pt Baskerville
by Photo·Graphics, Honiton, Devon
Printed in Great Britain by TJ Press Ltd, Padstow, Cornwall

Contents

Foreword

It is a pleasure as President of the Association for Science Education to write a preface to this thought-provoking book. The History and Philosophy of Science and Technology, and their applications throughout the existence of humanity, have great potential for interesting young people if taught well. It is also a valuable bridge across the so-called 'cultural divide' between the humanities and the sciences, which is an increasingly out-of-date and untenable concept.

Children can think of science and its applications as cold, hard, factual, and impersonal. That is plainly not true. From the day Archimedes leapt out of his bath shouting 'Eureka!', it is clear that scientific discoveries and the inventions based on them have been matters of great excitement to the scientists and engineers involved. Most of those inventions serve humanity in the form of bridges, pure water supply, electricity, and gas available in our homes, aeroplanes, trains, cars, hospital equipment — the list is infinite. Without them our lives would be primitive, dark, unhealthy, and poverty-stricken.

This book will awaken the interests of potential teachers, whether historians or scientists, and if taught well the subject will capture the imagination of young people of all abilities as relevant to their everyday lives, interesting, and at times very exciting. I commend it to you all, and hope it contributes to the broadening of education to the future benefit of all.

Platt of Writtle

Introduction

'It's all very well, but what use is it in the classroom?'

'I've got enough on my plate as it is without including a dollop of that!'

'I'm a convert, but please – can you provide the backup teachers need?'

'I love the subject, but how come you university lot make such a mess of it?'

These are some of the responses science teachers have offered when we have mentioned the history of science and technology to them over the past few years. As two members of the 'university lot', who firmly believe the subject has a place – or rather can fight for a place – in schools, we have found the comments most helpful. More exactly, we found them first of all depressing, then provoking, then finally inspiring. They prompted and encouraged us to produce this book.

In the following pages we have assembled a collection of essays which we hope goes some way towards meeting the genuine and perfectly understandable reservations teachers have about the history of science and technology. They are quite right to ask what role the subject might have in the classroom, and to wonder how it could ever be incorporated into a syllabus that is already bursting at the seams. They are right to turn to the community of professional historians for guidance and help with sources and resources. And, we confess, they are also right to wonder how we have managed the transmutation of gold into base metal – for too many teachers the excitement and verve of history of science seems to disappear once academics get to work on it. Thrilling stories of discovery and exploration, of victory against the odds, of controversy and conflict end up buried in scholarly articles, with their page-long footnotes, irrecoverable terminology, and esoteric points of reference.

The challenge facing historians of science and technology is to make their subject live and breathe in the school, without in the process sacrificing fact at the altar of classroom convenience. We hope in this book to have provided teachers with background materials, with actual examples, and with resources sufficient to enable them to cultivate a working interest in historical approaches to science teaching. Some teachers of history and science are already enthusiastic about the history of science; we expect that they will find the perspectives and practical examples developed by our contributors helpful in developing their enthusiasm, and in extending it into the classroom. In this introduction we will try to meet the commonest reactions of teachers to the subject by discussing recent changes in the teaching of history and science, by describing developments in the history of science over the past two decades, and by highlighting the main themes of the book.

There have recently been considerable changes in the way in which both history and science are taught in schools. Gone are the days when students were expected to concentrate on learning, by rote, a hard core of facts dictated by a 'content-based' syllabus. Today they are encouraged to acquire a broad range of critical and practical skills that will enable them develop a personal interest in the most general issues raised by a particular discipline, and to undertake research which is socially and culturally relevant to their own lives.

These changes have raised important questions about the materials appropriate in developing such skills in pupils and, as during previous periods of educational reform and innovation (particularly in the sciences), the role of the history of science in the school curriculum has come in for considerable discussion. Many teachers have sensed that this is a subject which offers a potentially powerful resource for their classroom work. Understandably enough, they have turned to the academic community of professional historians and philosophers of science for advice and guidance. Truth to tell, those teachers who have taken the trouble have met with hindrance as often as help. This has not arisen from a lack of enthusiasm on the part of historians of science, but instead because of the way in which their own practice has evolved. The comments with which we opened this introduction illustrate the parameters well enough, but we should like to identify in a little more detail three outstanding features of the history and philosophy of science which are commonly cited as impeding better links between school teachers and academics.

First of all, professional history of science as an academic discipline has developed and diversified considerably over the past twenty years. The teacher seeking to understand the 'state of the art' in a particular field of research is not unlikely to become as confused as many historians of science are themselves. It seems as if what historians write one year about, say Mendel's experiments with peas or Galileo's experiments on motion, is likely to be revised a year later, reassessed a year after that, and then re-evaluated once a little time has passed. The lonely historian in Kingsley Amis's campus novel *Lucky Jim* had the habit of answering his telephone at work with the hopeful line 'History speaking!'. Today's historians of science are more aware than ever before that if history speaks at all it does so, as often as not, in a babble. Different historians listen to and hear different things, and as a result what teachers find when they turn to them is as likely to be controversy and debate as it is to be consensus. This is a sign of the good health of the profession, but understandably off-putting to those who are looking for easily assimilated perspectives and straightforward answers to historical questions.

As if this were not enough, the teacher who turns to the extensive literature on the history of science finds that the many different perspectives and approaches represented there are clothed in the body-armour of jargon whose outstanding feature is a relish for the -ism and -ology. Academics, teachers must feel, are using language as Red Indians were wont to use war-paint. The simple truth is that professional historians of science have for the most part spoken in an idiom familiar and accessible to other professionals, with little regard for a wider audience. The work that results is quite inappropriate for the vast majority of practising teachers.

This is not to say that well-written, accessible studies in the history of science and technology do not exist. An admittedly small number of books and articles are available but these, alongside primary sources which have a potential use in school project work, are difficult to locate.

We have divided this book into three sections: 'Perspectives', 'Practice', and 'Sources and Resources', each of which deals with these problems of access and comprehension. For the remainder of this introduction, we should like to discuss in more detail the relevance of the history of science to the school curriculum and the ways in which contributors themselves address this and other issues of current interest.

A Changing Landscape

The concern expressed over school history and school science during the past twenty years reflects the increasing public awareness of the role of science and technology in our culture. Many of the most important and controversial issues of the 1980s, such as the arms race, nuclear power, pollution, high technology medicine, and genetic engineering, have focused attention on the relationship between science and society. Indeed, public engagement in debates about these issues has helped to break down the notion that 'science' and 'society' exist in separate, easily identifiable realms. Many people are now far less willing to leave what were once thought of as purely scientific or technological questions solely to that conventional figure of wisdom and trustworthiness – the scientific expert.

Teachers have found that the increasing attention given to the social and cultural role of science has not been adequately represented in school science classes. They also feel that students should be taught science in a way that enables them to develop balanced and informed opinions about the important debates concerning the interactions of science, technology and society, as they arise. Such concerns have raised fundamental questions about the content and objectivity of science teaching, and have brought teachers face to face with a dilemma. On the one hand teachers are anxious that students should be made aware of the role of science and technology in the development of twentieth-century society and culture. They also believe that students should grasp something of the detrimental as well as the beneficial aspects of this role. On the other hand, they are rightly concerned that students should not be given the impression that the study of science is intrinsically bad, or that scientists as experts are never to be trusted. They have fought instead to foster in students the ability to cope with science in a way which is both respectful of scientists' considerable technical expertise and wary of their fallibilities and weaknesses.

Many of these ideals have been adopted as official 'aims' of the new GCSE examination, the introduction of which marks the most radical change in secondary education in recent years. Furthermore, with dramatic changes in the 'A'-level examination likely to follow shortly – changes which have at least in part been designed to bring the final two years of secondary education into line with GCSE – we may expect that this new approach to science education will shortly underpin the whole of secondary education for students in the 14–18 age range. Let us therefore consider some of the proposed changes in a little more detail.

Broadly speaking, the most general aims of the history and the science syllabuses, as laid down by the GCSE *National Criteria* documents prepared by the Department of Education and Science, have more in common than ever before. Each insists that the course material should be accessible to *all* pupils regardless of ability; that it should, wherever possible, be relevant to their own social and cultural circumstances; and that it should provide them with a sound basis from which to develop an active interest not just in history and science, but also in those public debates which have either a historical or scientific dimension. Students, in short, should be made historically and scientifically literate.

In history, for example, students are expected to acquire the following skills: to understand the development of social and cultural values over time, and, when appropriate, make links with the present; to appreciate that a knowledge of history is rooted in an understanding of the nature and use of historical evidence; to develop the ability to locate and extract information from primary and secondary sources and to form this information into a logical argument. The emphasis has clearly shifted away from the mere acquisition of names and appropriate dates. The student today is being encouraged to consider the very nature of the historical enterprise by becoming practically involved in the production of historical knowledge. The rationale is that in this way the future citizen will be far better equipped to evaluate arguments which make appeal to historical evidence – arguments employed with alarming frequency by all those who seek to influence public opinion – by relating the claims and interests of writers or speakers to the way in which they have used the historical record.

Just as school history is no longer a chronology of dates, so school science is no longer a collection of facts. The teacher can no longer concentrate solely on laying the foundations for the practical and theoretical skills useful for the future professional scientist. Relatively few students go on to become scientists. Students are now receiving a scientific education which enables them to develop and maintain an active interest in matters of scientific import throughout their lives. More specifically, science students are being made aware that the study and practice of science are cooperative and accumulative activities subject to social, economic, technological, ethical, and cultural influences and limitations.

It naturally remains to be seen just how these ambitious aims of the new syllabuses are going to be interpreted in practice, but they would seem at the very least to necessitate the introduction of a new component into the science syllabus. This volume presents evidence that the history of science could provide just such a component, though we fully realize that much more work will need to be undertaken to make a convincing case. In helping to mount such a case, we think that it will be useful if we review some of the developments history of science has undergone over the past few years. These have been crucial in establishing the present-day character and image of the subject.

The Development of History of Science in Britain

Prior to World War II, research in the history of science was carried out mainly by amateur historians, often practising or retired scientists or science teachers. But with the expansion of tertiary education during the 1950s and 1960s, a number of university departments were set up for the development of what was often called 'The History and Philosophy of Science' (HPS). This rapid professionalization and expansion of HPS exerted a powerful influence on the kind of work which many historians of science produced during subsequent years. It also helped to generate some of the problems which now confront the teacher seeking advice on how to teach the history of science in schools.

The British Society for the History of Science (BSHS) is, for example the largest organization in Britain for anyone with a serious interest in the field, but it has been built up and run in the main by professional academics. As a consequence it reflects the interests and priorities of the professional rather than the amateur. The meetings organized by the Society offer an occasion for 'scholar to speak unto scholar', though they are often attractive enough to engage teachers, research students, and other groups. But it is only recently that the BSHS has become fully aware of the interest in history of science beyond the traditional sites of academic research. Although a number of teachers have produced fine work in the history and philosophy of science, an Education Section of the Society was established only in the past two years. It has had a considerable impact, partly as a result of its activities at the 1986 and 1987 annual meetings of the Association for Science Education, and has planned a number of events for the future. The publication of this book represents one of the first concrete results of the Education Section's work.

Such a volume has seemed all the more necessary since there now exists a wide range of specialist journals for the history of science, but all too few which are accessible to the non-expert. The teacher who browses through the articles in these journals looking for material which might be of use in the school classroom may well find it difficult even to locate the central problems being addressed by the authors.

The amalgamation of the history and the philosophy of science in the creation of HPS expressed the contemporary perception of the history of science as a constitutive part of the history of ideas. Many philosophers interested in scientific method and the status of scientific knowledge saw the history of science as a source of useful empirical evidence from which to draw support for their own theories. Much of the history written during this early period was consequently based upon one or other of the then popular philosophical models of scientific development, and therefore tended to focus upon the internal progress of the most general theories of the physical sciences. The names of some of the most famous recent philosophers of science, such as Karl Popper, Imre Lakatos, and Paul Feyerabend have therefore come to be closely associated with the history of science.

During the 1970s and 1980s many historians of science turned away from this traditional approach, and began to explore new areas of research and adopt new methods of historical interpretation. They sought to integrate their research with that being undertaken by historians of business, economics, theology, and politics, and drew upon methodological techniques developed within the disciplines of philosophy, sociology, art history, literary theory, and psychoanalysis – to name but a few. As a result, the very definition 'history of science' has expanded considerably, and historians of science have taken an active part in the investigation of the development of scientific institutions, the relationship between science and society, the social construction of scientific knowledge, the psychology of the scientist, and the popular role of science and science fiction in modern culture. Furthermore, the history of science as a discipline has become differentiated – some would say fragmented – into well-defined areas of specialization. Most professionals now see themselves not simply as historians of science, but as historians of medicine, technology, chemistry, physics, mathematics, and so on.

These are exciting developments which few could have anticipated 20 years ago. They are also challenging developments which force us to think afresh about the character, constitution and impact of science and technology through time. Some of the old tales of 'man's mastery over nature' or 'the machine as the answer to all problems' will need to be dispensed with, and there is bound to be some disappointment in the process. But there are rewards too, for the new history of science is quite as rich and multicoloured as many of its predecessors, and some would say much more so.

Unfortunately, some teachers may feel annoyed that the history of science remains a difficult and demanding subject. Yet a moment's thought reveals that this is so for any mature, but still growing, field of activity. Few teachers would expect to open up the latest issue of a journal of say, physics, and find there easily assimilable and predigested morsels to feed to their students. The same must be said about any of the articles which appear in a journal like *The British Journal for the History of Science*. But this does not mean that bridges cannot be built between the needs and interests of teachers and the work of academic historians. Lines of communication can be opened up, meetings and conferences can be run for both communities, and, with sufficient energy and commitment, a stream of usable resources can be produced. We certainly hope that school science and history teachers will find more common ground with professional historians of science over the coming years than they have in the past; and we hope, in addition, that this collection will help in this process.

During the last twenty years, historians of science have played an important role in fostering new ways of thinking about the nature of science and its relationship with the culture of which it is an integral part. The result is a mix of diversity and complexity, which is a force of strength rather than a weakness. We live today in a society which could not function without the products of science and technology, products which are now so commonplace as to have become largely invisible. Similarly, the concepts, language, and ideals of that science and technology have now become part of the very consciousness of our culture. As human beings immersed in that culture we are hardly aware of our physical and intellectual reliance upon science and technology, except perhaps until they fail us in some way, and it is impossible for us to separate off the scientific and technological components of our society and regard them objectively.

The history of science, however, provides us with many new perspectives both on science itself and on its role within society. For example, it focuses our attention upon societies within which the relationship between science, technology, and society is quite different from that of the present day. Our own society begins then to acquire a quality of strangeness which invites us to question relationships which were previously taken for granted. By studying the recent history of science and technology we can gain insights into the way in which a consensus has gradually been reached about many of the assumptions which we now take fully for granted.

All such perspectives will, of course, be delimited both by this kind of problem formulated by the historian, and by the methodological tools which are adopted in providing answers. Here again the variety of perspectives to be found within the history of science helps to prevent us from adopting the naive view that the study of history offers straightforward objective solutions to the problems faced by our society. Indeed the teacher will discover that the history of science is an extremely broad church within which the historical record is exploited to many different ends.

Many historians remain interested in the origins of present day science, and continue to trace its roots in earlier cultures. Others have abandoned modern science as a useful tool in investigating the science or natural philosophy of previous centuries, and seek only to employ the terms and definitions which were meaningful within the society under investigation. Similarly, different historians have recognized different reasons for engaging in research in the history of science.

Some see the history of science as an adjunct to science itself, and regard it partly as a method of popularizing and enhancing the public image of science and of scientists. At the other end of the spectrum, other historians believe that the most important role of the history of science is to provide empirical evidence with which to undermine the accounts of the status of scientific knowledge and the scientific method given by philosophers and scientists themselves. In practice, the majority of historians hold rather more sophisticated opinions which lie somewhere between these two extremes, but it is as well for newcomers to appreciate, from the start, that they will encounter such a rich diversity of opinion, and to know why this is so.

School Situations

How, then, is the history of science relevant to the teaching of history and science in schools?

In the past, history teachers have tended to fight shy of the history of science because of its concentration on the internal development of scientific theories; this seemed to make it irrelevant to other aspects of the general history course. Moreover, internalist history of science with its concentration on the details of theory development tended to assume a fairly sophisticated knowledge of science from the reader. More recently, however, historians have begun to appreciate and explore the immense importance of the history of science, technology, and medicine to the development of twentieth-century western culture, and there now exists a range of studies on this topic which demands little or no expert knowledge of science, and which is closely integrated with the general history of the period. If historians are keen to introduce their students to historical material which is relevant to their everyday lives and which often affords a wealth of local primary source material ideal for project work, the recent history of science, we believe, is a strong candidate.

The role of the history of science in science teaching is rather less straightforward. Many science teachers already include a small component of history of science in class and if, as is often the case, they are merely looking for interesting historical details to add to their science lectures there already exist some general histories of science which they can consult. Until recently a small number of science teachers were able to offer their sixth form students a very traditional AO-level course in the history and philosophy of science but, despite hopes that this course might be retained in some modified form at A-level, it now seems that it is likely to disappear with the phasing out of AO-levels. The most important role which the history of science might play in future science teaching, however, is in fulfilling some of the new aims of the GCSE science syllabus which were outlined above.

If all students are to become scientifically literate, in the sense that they are able to develop informed opinions on the important debates involving science, they will need to learn to evaluate the input into the debate made by all the important groups and individuals – including, of course, scientists. But being scientifically literate cannot mean having the ability to evaluate the claims made, advice given, or evidence offered by a scientific expert, since few professional scientists could manage this. The student can, however, be made aware of the function that these claims, advice, or evidence will typically play in the decision-making process, and one of the best ways of illustrating this process is by drawing on historical examples. These examples have the virtue of being, on the whole, much less controversial than current debates over the role of science and society and are, in any case, more conclusive in that the outcome of particular decisions and disputes can be readily evaluated.

The introduction of historical case studies into science teaching would also help to put important human elements back into the portrayal of science in schools. Many students, especially girls, find traditional science courses extremely dry because they present science not as a human achievement, but as a dull catalogue of facts and theories. Likewise scientists are all too often depicted as faultless automatons, hardly capable of error and with mysterious and incomprehensible insight into the workings of the physical world. Historical case studies which draw attention to the very human pressures under which most scientists have to work, the failures and disappointments which often follow long years of work, and to the communal effort which goes into the production of new scientific knowledge, would do much to disperse the daunting image of the infallible scientist which turns many students away from the study of science.

Perspectives

We have said enough, we hope, to illustrate the range of issues addressed by today's historians of science and to introduce some of the possible uses of the history of science and technology in the classroom. These are subjects dealt with in the opening section of the volume which is devoted to two major themes: the recent development and current state of the history of science as a discipline, and past and present ideas on the role of the history of science in the school curriculum. Opening this section, the chapters by Edgar Jenkins and Bill Brock discuss the long and fascinating history of attempts by educationalists and teachers to inject some history of science into science teaching, and invite us to consider how the history of science has actually become more relevant to the school curriculum during recent years. This is a theme taken up by Joan Solomon in the third essay, in which she takes a very hard-nosed look at why we should find space for the history of science in already overloaded science syllabuses. She casts considerable doubt on many of the most popular arguments which have been advanced in the past in support of the teaching of the history science, and draws attention to the severe practical problems which exist in trying to make the history of science relevant, and comprehensible, to a wide range of average-ability children.

Stephen Brush lends a transatlantic perspective to the debate by discussing the way in which the history of science has been used in physics teaching in the United States. The history of science has been used there as an aid to physics teaching for more than twenty years now, and Brush draws on a number of studies which were designed to assess the effect which this had upon the students' knowledge and perception of physics. He concludes that the history of science has had a considerable and beneficial effect on physics students, without reducing their ability to absorb the technical content of the science syllabus.

The last two essays of this section are devoted to a discussion of recent trends in the history of science. Harry Collins and Steven Shapin discuss the insights which some historians have gained by treating scientific knowledge not simply as the factual representation of the physical world, but as something which is socially constructed by the scientific community. Furthermore, they argue that all students would gain from being taught science in this way. Those who go on to study science at university would be better prepared to undertake original research, whilst those who merely take an interest in public debates within which science plays a role would not be perplexed by the spectacle of scientific experts failing to agree about scientific facts.

Finally, Richard Jones outlines changing views of the historiography of science throughout the twentieth century, and shows how changes in philosophy and sociology have been reflected in the kinds of account of the scientific past produced by historians. He concludes by discussing some of the fundamental historigraphical problems which have been generated by the abandonment of the traditional notion of the objectivity of scientific knowledge. This article, we think, provides a useful overview of some of the most fascinating theoretical problems currently faced by historians of science.

Practice

The second section of the book looks at the practical application of the history of science in the school classroom, and includes a number of case studies. In the first essay of this section Roger Hennessey looks from the point of view of a senior inspector for history at some of the reasons why the history of science should be included in the history curriculum. This theme is further developed by Joe Scott, author of the textbooks for the courses *Medicine Through Time*, and *Energy Through Time*, and in his essay he looks at history of science in the Schools History Project.

Piyo Rattansi then considers a problem which can often arise when teaching science to a class of students of mixed cultural background. He suggests that by presenting western science as the most fundamental and objective account that can be given of the universe, teachers may, however unwittingly, devalue the parental or ancestral cultures of students from non-European countries. Rattansi argues that teaching the history of science could overcome this problem by making children aware that scientific concepts have changed and evolved as our culture has developed, and are not simply the result of the application of some 'scientific method', which is the sole preserve of the western world.

The next four chapters present practical examples of the way in which the history of science can be used in the teaching of physics and chemistry. David Gooding discusses the experimental work of Michael Faraday, and suggests how it might be used to inform children about the way in which great scientists proceed when working on 'frontier science'. The detailed laboratory notebooks kept by Faraday whilst working towards some of his most celebrated experiments present a unique record of the discovery process in action. They could become an effective resource in showing children that important discoveries are not the result of unique 'eureka' experiences, but are rather the culmination of months – and sometimes years – of painstaking research and effort.

The paper by Stephen Pumfrey should be of particular interest to many teachers as it represents, as far as we are aware, the first attempt to present children with an explicitly 'constructivist' account of a well-known episode in the history of chemistry. Pumfrey shows how the approach to scientific knowledge and practice outlined by Harry Collins and Steven Shapin throws new light upon the problem of the 'discovery of oxygen', and discusses the important role which this kind of case study might play in future science teaching. Peter Ellis also discusses the role of the history of chemistry in chemistry teaching, and emphasizes its particular relevance to the practical requirements of the new GCSE chemistry examination. His work has the considerable merit of being based upon many years of experience in the use of history in science teaching, and he has reproduced part of one of his course units at the end of his chapter.

The role of the history of electricity in physics teaching is discussed in the essay by John Roche, who argues forcefully that physics teachers should not use inaccurate or made-up history as a pedagogical aid when teaching physics. He insists instead that the actual history of physics is, in any case, more useful than any idealized account. Roche believes that by studying the history of important concepts in physics we become better able to explain them consistantly, and illustrates his thesis by discussing the concept of 'electrical potential'.

This section of the book concludes with a chapter by Edward Yoxen on the controversial topic of scientific fraud. Yoxen discusses some of the fascinating studies which have been carried out recently by historians of science on what he terms 'soft fraud', in which scientists 'clean up' or selectively publish experimental results in order that they should better fit the claims being made. Yoxen also discusses a board game which he has devised called 'The Fraud Game' in which players take the part of scientists trying to gain credibility by publishing scientific papers. The papers may represent either real research, for which the player has, of course, to expend his finite resources, or fraudulent research which is cheap, but which carries severe penalties if detected. The game represents a novel way of introducing students to some of the pressures and temptations under which many scientists have, in practice, to work.

Sources and Resources

The four contributions which make up the final section of the book are intended as a guide to help teachers to locate the advice, secondary literature, and primary source material which they will need in order to design and implement school projects in the history of science. David Knight discusses scientific textbooks, and explains how they can provide a readily accessible primary resource for school projects. The scientific textbooks of the past illustrate the scientific concepts which earlier generations of scientists took for granted, and even textbooks from the early decades of the present century may be usefully contrasted with those in current use.

Paul Weindling takes a look at the many journals which now carry articles related to the history of science, and the various societies which teachers may find helpful when setting up project work. There are now well over two hundred such journals in history of science and medicine alone, and with an increasing number of scientific societies and institutions beginning either to engage in, or else to support, research in these fields, we hope that teachers will find this kind of survey article welcome.

Local institutions represent an important source of help and advice for teachers interested in the history of science, and Stella Butler takes up this theme by discussing the role which might be played by local science centres, museums of industry, and museums of science and technology. Most of these institutions now offer rather more than just a static display of objects, and many have comprehensive education programmes. Any teacher who is fortunate enough to live in the vicinity of either a science centre or a museum (a list is included at the end of the article) should find that a visit will save a great deal of time and effort when it comes to setting up locally based project work for his or her students.

Many teachers regularly use films and videos as part of their science or history lessons, but most will probably not be aware of the large number of films which have been made on the history of science. In the final essay of this volume Cathy Grant discusses the range of films and videos on the history of science which can be hired for school use, and concludes with a comprehensive list of titles.

Whither the History of Science in Schools?

This book represents the first book-length attempt in Britain to make teachers more aware of the rich educational potential of the history of science, and as such we hope that it will find a place in many classrooms. We have tried to represent what we believe are the most important trends both in the academic study of the history of science, and in its application in school science and history. But we are only too well aware of the limitations of our enterprise. Any book which seeks to represent a broad cross-section of opinion must do so at the expense of exploring any particular approach in depth, and this means that a great deal more work will be required in the future.

Taken individually, the following essays express a wide range of perspectives on the history of science and convey different hopes and aspirations for its role as a subject and as an approach in schools. Taken together, a more consistent impression should emerge; an impression of a field of study which is rich in insight into our past, pertinent to the present, and full of interest for the future. We hope that we have managed to convey this in the book as a whole. A time may come when the kinds of complaint with which we opened this introduction will only rarely be heard. This is something we greatly look forward to, and if *Teaching the History of Science* helps bring that time about, it will have served its purpose well.

Acknowledgements

Most of the essays which appear in this book were presented originally at a conference on 'The History of Science and Technology in the School Curriculum', held in Oxford in 1987. Michael Shortland, who organized the meeting, would like to thank all those who tóok part in the event and contributed to the book. The meeting received generous help and support from the Department of External Studies at the University of Oxford, the Oxford Schools' Science and Technology Centre, the Standing Conference on Schools' Science and Technology, the Salter's Trust, and the Institute of Physics (Education and History groups). In helping to make a volume from the papers given at the Oxford meeting, we would like to thank the members of the Education Section who gave us support and colleagues who gave us support and advice, particularly Elspeth Crawford, Bob Olby, Mari Williams, and John Roche. We are most grateful to the Baroness Platt who kindly agreed to write a foreword for the book.

Last but by no means least, one organization has been constant in its moral encouragement and financial support both for the original meeting and for this publication – the Oxford Trust. It is literally true that without the help of the Trust, and particularly its director Paul Bradstock, none of our own work for this project would have been possible. Our thanks to the Trust are correspondingly great.

Michael Shortland
Andrew Warwick

Perspectives

Why the History of Science?

Edgar Jenkins

When the Duke of Argyll delivered his Presidential Address to the British Association for the Advancement of Science meeting at Glasgow in 1855, he told his audience that 'What we want in the teaching of the young, is, not so much the mere results, as the *methods* and, above all, the *history* of science . . . that is what we ought to teach, if we desire to see education, well-conducted to the great ends in view'. The Duke was, of course, speaking at a time when the British Association was particularly active in the schooling of science and his reasons for advocating the history of science in schools so forcefully are, therefore, of some interest. 'It is not', he claimed, 'merely for the sake of investing the abstractions of science with something of a living and human interest', nor 'merely to impress . . . results better on the memory'. It was, as might be expected given the concerns of the British Association in the mid-nineteenth century, for the benefit of science itself. By helping students to 'better appreciate the labour of others', they would be better able to recognize 'the golden strands of truth in the midst of much error'.[1]

When the same Association addressed itself to school science education in the particular and significant circumstances of the First World War, it produced a report which included a section entitled *Human Aspects of Science*. The report urged that there be 'more of the spirit and less of the valley of dry bones' in school science teaching and suggested that one way of achieving this would be by teaching 'lessons on the history of science'. History and biography, it was claimed, enabled a comprehensive view of science to be constructed 'which would not be obtained by laboratory work' and they 'supplied a solvent of that artificial barrier between literary studies and science' which a school time table usually sets up.[2]

The British Association report of 1917 had one feature of particular interest. Addressing the question whether there were any general principles to guide a teacher in structuring a curriculum to meet the needs of pupils, the British Association committee which prepared the report sought criteria to distinguish 'things suitable' from 'things unsuitable' for pupils of different ages. The committee, drawing upon the ideas of T.P. Nunn, identified three 'motives' which had characterized the scientific endeavour at different periods in its history and convinced itself that these motives, *wonder, utility*, and *systematization*, were also present to different degrees in the minds of pupils at different stages of their conceptual development.

Nunn's parallelism invoked the history of science, as he perceived it, to constitute a learning theory which could be used to determine both the sequencing of topics within a school course and the strategy whereby they should be taught. Nature study, for example, was appropriate for younger pupils because it appealed to the *wonder motive* dominant at that age. Similarly, Archimedes' principle should be taught to boys and girls of an age when the *utility motive* was to the fore, and introduced not simply as a property of fluids but as a means of understanding why ships are able to float.[3] It was an elegant, convenient, and functional parallel between the history of science and intellectual development and one which survives to retain a degree of contemporary significance. For example, the July 1987 issue of the *Journal of Chemical Education* contains an article entitled *Parallels between Adolescents' Conception of Gases and the History of Chemistry*.[4] The parallel drawn in the article is between Aristotelian conceptions of a gas on the one hand and, on the other, the shift in understanding of the nature of the gaseous state in the developing mind of the pupil. Clearly, some elements of recapitulation theory are not too far removed from the thinking of at least some constructivist psychologists, for whom the manner in which students construct meaning is of central importance. However, as Joan Solomon points out in this volume, children's ideas are not theories consistently applied, and desired cognitive changes are not necessarily brought about simply by presenting pupils with the arguments in a past scientific revolution.

Examples of benevolence (not always of an unqualified kind) towards the educational merits of the history of science can be found in a very wide range of publications published throughout this century and in many countries of the industrialized Western world. It is unnecessary to cite further evidence in support of this view, but attention should perhaps be drawn to the position taken by the Association for Science Education in its Consultative Document, issued in 1979 and entitled *Alternatives for Science Education.* A course in the history and philosophy of scientific ideas was proposed as one of a number of options to be followed by pupils in years 3–5 of secondary education. Its function was to act as a valuable bridge between science and the arts and the social sciences, either as a course of study in its own right or as a contextual course taken in conjunction with conventional science courses. It was, however, a proposal which did not survive the processes of consultation. The reaction of the (now) Royal Society of Chemistry can be regarded as representative of the view which came to prevail, namely that 'The history and philosophy of science are often best appreciated after the pursuit of various science disciplines over a period of time . . . There appears to be a strong case for weaving in such facets of science education when and where appropriate, rather than presenting them as separate entities'. Given this reaction, the recent initiatives of the Association for Science Education and the British Society for the History of Science to encourage the teaching of the history of science and of technology in schools are particularly welcome.

Claims to accommodate the history of science within the school curriculum have a long history; much longer, it may be noted, than the period during which the state has undertaken to fund secondary education. This book, with its aim of encouraging dialogue between professional historians of science, science teachers, history teachers, and others (of whom I am one) is part of that history and it offers a timely opportunity to reconsider a number of issues and to review strategies, tactics and resources. The history of science and the social context of science itself and of science education have all been transformed during the past quarter of a century or so. In addition, schools face a period of central control over the curricula of maintained secondary schools which many had thought vanished for ever, and the history of science as an academic discipline has to contend with the processes of 'rationalization' and contraction amid the chilly, not to say hostile, climate which currently characterizes higher education.

Before returning to address more directly the question 'Why the history of science?', it may be helpful to take stock of the present position. Cataloguing the flora in the secret garden of the curriculum of secondary schools in this country is a notoriously difficult task. It is, however, clear that much is being done and that there are individuals, groups, organizations, and institutions willing to help in an attempt to achieve more. There are some growth points, notably in courses concerned with science, technology, and society, such as SISCON in schools, the ASE science in society project and the SATIS 16–19 initiative announced early in 1987. The history of science finds a home in some of the Nuffield science courses, sometimes on an optional basis. Courses of sixth form general studies, many of them containing material concerned with the historical and philosophical aspects of science and technology, continue to flourish. In 1986, for example, the A-level examination in General Studies conducted by the Joint Matriculation Board attracted about 40 000 candidates, far more than the entry in any other subject at that level. There are also the important components of the history of medicine and of man's use of energy in the Schools History Project, which are discussed later in this book.

GCSE courses, governed by the General Criteria for GCSE examinations, seem to offer something of a new opportunity, although it is difficult not to be sceptical about how a desideratum such as 'the promotion of an awareness that the study and practice of science are cooperative and cumulative activities . . . subject to social, economic, technological, ethical and cultural influences and limitations', might be translated into practice in GCSE courses in, for example, physics and chemistry.

Other new opportunities to accommodate the history of science within schools might stem from developments at AS level in history, the sciences, and the history of science; from modular approaches to curriculum construction, and from the Technical and Vocational Education Initiative. On the debit side, science text books and other materials intended for use by pupils seem to present an extraordinarily naive and positivistic view of science and to make much less reference, both in the text and by way of illustration, to the history of science and technology than was the case forty or more years ago. An analysis of CSE, O- and A-level text books in physics, chemistry, and biology, undertaken a couple of years ago by Douglas Newton, makes discouraging reading.[5] His conclusion, after analysing sixty recent school science texts, was that the books were generally dehumanized and impersonal, and unsuitable for use by teachers. Also disappointing is the failure of the Department of Education and Science policy statement, *Science 5–16*, to do more than assert that 'the social and economic implications of scientific and technological activity . . . have a place in science education, always provided that the teaching is essentially concerned to foster an education in science itself'. There is clearly no encouragement for historians of science in the further comment in the policy statement that 'too many examples of . . . applications of science . . . reflect past developments rather than current ones'.[6] My disappointment at the policy statement extends to the recommendations of the *Secondary Science Curriculum Review* which, at one stage, had seemed to promise more than it has now delivered, although some papers of direct interest to historians of science have still to be published. The recommendations of the *Review* for 'Science for All', a title with a powerful historical resonance, include only two references to the history of science and technology. The first relates to the need to reduce the contents of science syllabuses to ensure that the economic, environmental, political, and social implications of science receive adequate attention. The second calls the history of science and technology in aid of multi-cultural education, arguing, among much else, that school science should 'aim to engender in young people the realization that progress in science has never been the sole prerogative of the Western European mind' and that 'human beings in the past often made scientific strides commensurate with the rate of progress familiar in the twentieth century'.[7]

While welcoming this aim, it is necessary to express the hope that it reflects much more than a token response to multiculturalism and to acknowledge that developing the required curriculum materials appropriate to pupils of a wide range of ability is a particularly challenging task. The current open-ended, discovery, 'process', or investigative approach to school science education supports and reinforces a long established and, for many years, discredited view of science which sets it apart from other disciplines in ways which too easily confirm the impression among pupils that the only wisdom is contemporary wisdom, the only valid truth is contemporary scientific truth and that earlier scientists and 'scientists' of other cultures simply 'got it wrong'. Indeed, much contemporary science teaching in schools must leave some pupils wondering how earlier investigators of the natural world can have been so blind to 'the obvious'.

This, of course, is a consequence of the assumption inherent in the official rhetoric and, it seems, in much of the practice of school science education that a given set of observations leads to inevitable conclusions. This assumption, in turn, involves a failure to recognize what has been called the metascience of scientific activity, i.e. the unidentified and tacit assumptions embedded in the social and communal dimensions of the scientific enterprise. Rattansi, writing in this volume, has exposed this point particularly well in his rewriting of a part of John Lewis'

Physics 11–13 to conform to the textbooks in physics which were being used at Cambridge when Newton was an undergraduate there in the 1660s. Rattansi has also identified the results of failing to make pupils aware of the conceptual framework within which scientific activity always takes place: a deeply negative image of non-Western cultures, and the dismissal of the giants of Sumeria, Babylon, Egypt, India, China and the Semitic and Hellenistic worlds, upon whose shoulders Newton acknowledged himself to stand.

A preceding paragraph has hinted at the difficulty of establishing with any confidence the extent to which the history of science and technology are taught in secondary schools and the manner in which they are accommodated. Even less, perhaps almost nothing, seems to be known about the quality of pupils' learning. In the United States, where there has been a somewhat greater interest than in this country in using the history of science to teach science – the *Harvard Project Physics* is perhaps the most familiar example – there is evidence that the inclusion of a historical dimension in science courses influences students' attitudes towards science itself and leads to an improvement in their understanding of the complexity of the scientific enterprise (see Stephen Brush's contribution to this volume).

In the United Kingdom, although there is reason to hope that many pupils enjoy the history of science, one can only guess that pupils' understanding of the history of science or technology is very modest and much less than it needs to be if they are to have a historical perspective upon the enterprises which have had, and continue to have, such astonishingly powerful and often profoundly subtle effects on all our lives. That guess is coloured by my experience of pupils and student teachers in schools and, rather more, of the poor quality of answers concerned with the history of science or technology from A-level candidates whose general intellectual ability, as judged by their performance on the rest of a General Studies examination, is rather above average. The weakness of these candidates is not their lack of analytical skills or the absence of a historical perspective, but a lack of basic knowledge and the presentation of ideas within the excessively positivistic tradition to which reference has already been made. Such ignorance, always discouraging, can of course occasionally be a source of amusement, as with the claim that 'Gregor Mendel was a monk who spent most of the nineteenth century in a convent doing experiments on breeding'!

Why therefore, the history of science? What are its claims to a place in school curricula? The history of these claims, referred to in the preceding paragraphs, suggests at least three broad answers to this question.

First, the history of science is seen as offering a 'humanizing' element in an otherwise dehumanizing scientific education. This was a claim pressed with particular vigour in the years immediately following the First World War and, more generally, during the inter-war period. This was, of course, a time when there were significant advances in the emancipation and recognition of the history of science as a scholarly discipline. However, it is possible that the advocacy involved something of a reaction to the prostitution of science, as some saw it, in the slaughter of 1914–18. Science by itself, and from the British perspective German science was the supreme example, was not a humanizing study. This, of course, was by no means a new criticism, but it was articulated with some clarity in 1916 by Sir Richard Livingstone in his *A Defence of Classical Education*. The fundamental weakness of science, as Livingstone saw it, was that it 'tells us hardly anything about man. The man who is our friend, enemy, kinsman, partner, colleague, with whom we live and have our business, who governs us or is governed by us, never comes within our view'.[8] If the proper study of mankind is man, then science, for Livingstone, was defective. The history of science, cast it should be noted in a subordinate and remedial role, seemed to offer something of an antidote to this lack of humanity. Livingstone's challenge to the educational aspirations of science produced a vigorous response, notably and predictably from H.G. Wells, but it has not been rebutted by argument. Indeed, any such rebuttal would seem to be difficult as long as science is presented in schools as a pre-professional course of study, with all that this implies.

A second, and somewhat different rationale for teaching the history of science was evident in this country (and elsewhere) after the end of the Second World War, and came to the fore in the years immediately preceeding the establishment of the Robbins committee on higher education which reported in 1963. Between 1955 and 1963, when the secondary schools were accommodating the consequences of the boom in the post-war birth rate and of the tendency of pupils to remain longer at school – consequences referred to fondly as the 'bulge' and the 'trend' respectively – competition for university places grew to become intense. The competition, it is of interest to note, was particularly severe in the case of admissions to undergraduate courses in pure science.

The response of the schools was the earlier and narrower specialization about which the Crowther report of 1959 had so much to say. It was the time of C.P. Snow's Rede Lecture on *The Two Cultures and the Scientific Revolution* and of the voluntary consortium of 'ABC' schools. The acronym, standing for agreement to broaden curriculum, was applied to a group of schools which gave a public undertaking not to devote less than a given percentage of their time to non-specialist studies. The period was marked by a lively debate about science as a general study in the sixth form. This was, for example, the title of a conference held in Oxford in 1959 at which the history of science figured prominently in the conference discussion. Its role, potential if not actual, was to straddle the 'Snow line', i.e. to constitute common ground between the arts and the science specialists.

This, of course, was the role which Herbert Butterfield anticipated in *The Origins of Modern Science*, wherein he claimed that 'it is hardly possible to doubt the importance which the history of science will sooner or later acquire both in its own right and as the bridge which has so long been needed between the Arts and the Sciences'. The history of science as a 'bridge' receives much sympathy in the report of the 1959 Oxford conference, but the arguments against it are also spelled out.[9] These arguments, which relate to such issues as teacher competence and the extent to which a study of the history of science presupposes a knowledge of science itself, seem to have lost little of their validity, and they serve as a reminder of the difficulties facing anyone pressing for an expansion in the teaching of the history of science in schools at the present time.

A third, and enduring, justification for teaching the history of science in schools has been the value of such teaching in providing students with a richer insight into the nature of science than that commonly provided or encouraged by conventional science teaching. In particular, a critical dimension can be added to scientific education, and science can be presented as part of a wider cultural heritage. When James Conant wrote the introduction to the *Harvard Case Histories in Experimental Science*, first published in 1948, he referred to the importance of giving students majoring in the humanities or the social sciences a feel for what he called 'the tactics and strategies of science'. Almost forty years later, philosophical and historiographical changes, the emergence of sociological perspectives, the transformation of the relationship between science, technology and society, and growing public concern about aspects of that relationship, mean that the tactics and strategies of science, as revealed in the *Harvard Case Histories*, require reinterpretation.

Illuminating science remains an important objective in general education but both the science and the nature of the illumination have changed. The history of science, broadly interpreted to embrace a variety of perspectives and approaches, seems to be admirably suited to make a major contribution to this end. John Ziman's account of the contribution which the history of science may be able to make to courses concerned with science, technology, and society suggests a way forward.[10] Acknowledging that the history of science is a 'lovely academic discipline' but warning that it is mainly for the *aficionado*, he describes the historical approach to studies in the field of science, technology, and society as having 'immense educational merits'. It offers, I believe, a means whereby issues of contemporary importance and real interest to many pupils can be explored. As always, translating those 'immense educational merits' into practical teaching schemes poses severe problems, most of them by no means new. It is nonetheless within the context of science, technology, and society courses, as well as of 'conventional' history courses, that much of interest and value is being done in the history of science and technology in schools. Some of the best work draws upon the local context and calls upon the resources of teachers qualified in different disciplines. However, as Brock notes in reviewing recent initiatives in this book, 'most of the experimentation with History of Science in the science curriculum has actually involved History of Technology, Industrial Archaeology (and perhaps History of Medicine) rather than the history of scientific ideas'.

It is possible to speculate about a rather different role for the history of science in school science education which is more radical and more overtly political than those touched upon so far. The speculation is whether the history of science can be used in a systematic and deliberate manner to force a reconsideration of some of the objectives of school science education itself. This, of course, is very different from using the history of science in a remedial capacity, either as a bridge between the two cultures or in providing the 'humanistic garnishing' which conventional science courses have been deemed to need. In these remedial contexts, the history of science has been called in aid of a curriculum in difficulty and its task has been that of buttressing curriculum objectives established without reference to the history of science itself.

The more radical contribution can be made clearer by reference to a distinctive aspect of secondary school science education in the United Kingdom: the practical teaching of science in the laboratory. The role of the teaching laboratory in secondary education was established in the second half of the nineteenth century, and despite a transformation of the social context of science and of scientific education, has undergone no fundamental change. The role has, of course, been expressed in different terms at different times and been supported by different educational psychologies. Armstrong's heurism, with its emphasis on the teaching of scientific method as understood by Spencer, J.S. Mill, and Thomas Huxley, relied heavily upon a faculty psychology for its support. When that psychology collapsed, so did the case for heurism, although other factors were also involved.

Today, the policy statement of the Department of Education and Science leaves no doubt as to why science is to be taught in schools, presenting its opinion in terms that Armstrong would have welcomed. The essential characteristic of an education in science is that it introduces pupils to the methods of science, and the value of a scientific education lies in the claim that 'Each of us needs to be able to bring a scientific approch to bear on the practical, social, economic and political issues of modern life'. The supporting psychology accepts that a pupil's knowledge must be 'personally constructed' and, in the case of science, the learning 'processes' of observation, inferring, hypothesizing, etc. show a remarkable similarity to the 'processes' whereby science itself is advanced. Indeed, for some cognitive psychologists, the two sets of 'processes' are identical. Bruner, for example, has asserted that 'intellectual activity everywhere is the same' and that the pupil learning physics *is* a physicist and can learn physics better this way than by doing something else.[11] However, it is not the intention in the present context to embark on a critique of the 'process approach' which characterizes the rhetoric of contemporary school science education and informs the practice of, for example, GCSE assessments in science.[12] Nor is it to ask whether historians are able to challenge such an approach and expose the deficiencies which some believe it to have. It is to wonder whether the history of science, broadly conceived to embrace a variety of perspectives, can offer an alternative view of the role of, for example, the laboratory (and of experimentation in particular) in the generation and accreditation of scientific knowledge. If it can, then the way is open to a reconsideration of the existing function of practical work in school science education which is not only long overdue, but increasingly urgent if the educational potential of science is to be tapped by the majority of pupils with greater success than hitherto.

Curiously, that educational potential is beginning to be explored and realized outside the classroom, notably by adults engaged in controversial issues with a scientific or technological dimension.[13] Wishing to challenge scientific experts and their advice, some adults have been forced to question the nature of scientific fact, observation, and theory and, thereby, to identify assumptions which, previously, they had not so much accepted as failed even to acknowledge. Alternative assumptions and alternative ways of thinking about the nature and purpose of the scientific enterprise are thus engendered. The history of science seems particularly well-suited to contribute to the generation of some of these alternatives within the school context. This, perhaps, is not an objective likely to be welcomed immediately and with enthusiasm by the professional scientific community, but it is nonetheless one which promises to enrich and empower some of the lay citizens of the twenty-first century currently in our schools.

According to Guerlac, the 'central business of the history of science is to reconstruct the story of the acquisition of knowledge of the physical and natural sciences and, perhaps, above all, to study science as a human activity, how it arose, developed, expanded, how it has influenced or been influenced by man's material, intellectual, even spiritual aspirations'.[14] It is a description of 'the business' which is attractive, and one which seems to imply a major role for the history of science, not only in contributing to, and shaping the objectives of, a humane scientific education, but also in suggesting a new dimension for the teaching of history in secondary schools.

NOTES

1 British Association for the Advancement of Science (1856), p. lxxxiii.
2 Ibid. (1918), p. 140.
3 Ibid. pp. 135–7.
4 Mas, Perez, and Harris (1987).
5 Newton (1986).
6 Department of Education and Science (1985), pp. 12–14.
7 Secondary Science Curriculum Review (1987), p. 4.
8 Livingstone (1916).
9 University of Oxford Institute of Education/Science Masters' Association (1960).
10 Ziman (1980).
11 Bruner (1960), p. 14.
12 Millar and Driver (1987).
13 Layton, Davey, and Jenkins (1986).
14 Guerlac (1977).

Past, Present, and Future

Bill Brock

In May 1986, the Education (Research) Committee of the Association for Science Education (ASE) set up a working party to examine ways in which 'science teaching in schools might be enriched, and science shown to be a cultural activity, through the use of the historical perspective on science and technology.' This development followed intensive discussions between the ASE, the Historical Association (the leading organization of teachers of History in British schools) and the British Society for the History of Science (BSHS). These discussions had in their turn been promoted by the ASE's 1979 discussion document *Alternatives for Science Education*, which had asserted categorically that history and philosophy of science in the Sixth Form would assist in broadening the science curriculum, and that science could be studied in a variety of contexts beyond 'science as science', including 'science as culture', 'science and citizenship', 'science at the workplace and in leisure', and 'science and survival'. Clearly, several of these contexts implied an historical dimension in school science. That was in 1979.

Although these views were a response to the social, political, and economic pressures of the 1980s, they were by no means new or original, for we find them present in an influential British Association for the Advancement of Science (BAAS) report on the secondary science school curriculum in 1917, and in the equally influential government report of 1918 entitled *Natural Science in Education* (the J.J. Thomson Report).[1] It was the latter which effectively established the British divide between general education to 16 and the specialized Sixth Form education from 16 to 18. Both the BAAS report on *Science Teaching in Secondary Schools* and the Thomson Report urged that:

It is desirable ... to introduce into the teaching some account of the main achievements of science and of the methods by which they have been attained. There should be more of the spirit, and less of the valley of dry bones ... One way of doing this is by lessons on the history of science.

Un peu d'histoire

Why this interest in the history of science in schools during the middle of the First World War? Although there had been some development of academic interest in history of science in France and Britain at the end of the nineteenth century, the First World War interest came more from school science teachers, many of whom had classical education backgrounds. (This was, of course, particularly true of science masters in private schools, who in 1901 had formed the Association of Public School Science Masters – the parent of today's Association for Science Education.) Such classically trained teachers were, no doubt, familiar with Matthew Arnold's debate with T.H. Huxley over classical and scientific values and knew that both Victorian sages had agreed that the definition of a cultured person was someone who knew about the best that had been thought and said in the past, and that the promotion of human culture involved both the classical and scientific cultures. As Arnold put it in *Literature and Science* (1885):

By knowing ancient Greece, I understand knowing her as the giver of ... the guide to a free and right use of reason and to scientific method, and the founder of our mathematics and physics and astronomy and biology ... By knowing [about] modern nations, I mean knowing also what has been done by such men as Copernicus, Galileo, Newton, Darwin.

Such Arnoldian arguments became particularly significant during the First World War when, because the utilitarian need for more scientific instruction in schools threatened to reduce the time available for classics, many classics teachers retaliated by accusing scientists (and hence science teachers) of being 'barbarians' and of teaching a 'dehumanizing' subject. It therefore became particularly important for science teachers to establish the defence that, despite gas warfare, and the successful fixation of nitrogen for making bigger and better explosives, science nevertheless possessed a cultural value and exerted a truly humanizing influence upon individuals who studied it. The history of science, through biography and stirring stories of scientific discovery and achievement in the way that Samuel Smiles had so successfully advertised the heroic qualities of Victorian engineering, would counter the dangerous arguments of the die-hard defenders of the classics like Richard Livingstone. Such cultural teaching of science also complemented the views of educators and philosophers of education who had introduced the ideas of Herbart into Edwardian teacher training colleges, that teaching was more effective if children's interest was captured.

Such Edwardian educationists were also much under the spell of the view that an individual child's mental development paralleled that of civilization. According to this phylogenic view of mankind's history, men had first studied nature out of wonderment; then because it was useful to do so; and finally for purposes of systematization and intellectual gratification. It followed that the teacher's first approach to science ought to seize upon the younger child's sense of wonder, some of which could be achieved through the history of science. This fitted, too, the feeling amongst most public school science teachers (whose classes were usually identified according to boys' ability at classical languages and not science) that most children were incapable of going much beyond the stages of wonder and utility; and consequently, that if science was to be taught to 'all' (as they sincerely believed) then a generalized science curriculum which focused upon the everyday and on the applications of science would have to be developed.

This 'Science for All' curriculum, which emerged between 1916 and 1920, also included elements of history of science.[2] (It is ironic that the ASE's present campaign which again uses the slogan 'science for all', does not recognize its historical roots.) Finally, we should note that the First World War advocates of history of science in schools were also concerned that too much emphasis had been placed, by Henry Armstrong and others, on teaching science through experiment, by discovery, or heuristically; and it was felt that awareness of the methods scientists used in their investigations might be more rapidly gleaned through the study of historical examples and through demonstration lessons.

During the 1920s and 1930s, then, there was a respectable and sustained interest in the history of science in British secondary school science teaching. It was seen as a way of breaking down premature specialization and helpful in the promotion of scientific understanding amongst children who were not going into scientific careers. It was also seen as a way of counteracting the criticism of classics teachers that the sciences were 'dehumanizing', and it was felt to be a valuable way of teaching the nature and methods of science. Last, but by no means least, it interested those philosophers of education like Richard Gregory and Percy Nunn who believed that the mental development of mankind was repeated in the mental development of individuals.

Consequently, many teachers' training colleges, notably at Liverpool and in London, offered courses in history of science for prospective science teachers, while, from 1921, University College London offered them a part-time evening M.Sc. course in the History and Methods of Science. Teachers' interest in the subject was also sustained by numerous articles in the *School Science Review* (founded in 1920) and by the prolific textbook writings of E.J. Holmyard, J.A. Cochrane, and J.R. Partington, all of whom wrote from an historical perspective. Moreover, when the two first professional journals of history of science appeared in 1936 (*Annals of Science*) and 1937 (*Ambix*), both were warmly welcomed by the *School Science Review* on the grounds that school teachers 'knew from experience the value of historical details in arousing and maintaining interest and in meeting the criticism that science is unhuman'. Both journals, it recommended, ought to be placed in every school library.[3]

Not surprisingly, therefore, when the British Society for the History of Science was founded in 1946, it included a number of former and practising schoolteachers among the founding membership. During its first five years the Society held several meetings with the then Science Masters' Association and Association of Women Science Teachers (the forerunners of today's Association for Science Education) in order to consider how the history of science could be further promoted in schools. Although there were some positive results of this post-war activity – BBC broadcasts to schools in 1950 and the creation of two sixth form examinations in history of science in 1952 shortly after the General Certificate of Education replaced the old School Certificate examinations – the overall effect was small. For, despite these developments, history of science remained an unprescribed and elective element in science teaching and the number of candidates actually taking O/A-level examinations was, and remained, a minute proportion of those taking examinations in chemistry, physics, and biology.[4] Although history of science teaching in British universities developed considerably during the 1950s and 1960s, the actual percentage of science teachers trained in history of science during the last thirty years has been extremely small.

Despite further encouragement for teachers to interest themselves in history of science – C.P. Snow's 'Two Cultures' debate in 1959, the emergence of 'General Studies' in the sixth form in the 1960s (which arguably actually diverted the growing interest in history of science), the importation from America of 'Science Case Histories' in the same decade, and the arrival of Nuffield O-level science syllabuses which included historical materials – it seems fair to generalize that until very recently history of science has played only a very minor role in school science teaching.

The explanation for this is clear: the ASE, which has a current membership of 17 000 teachers, had no positive policy on the role of history of science; the BSHS for its part had become deeply involved in the professionalization of history of science in the universities and polytechnics and simply ignored its school teaching constituency. Similarly, although the Nuffield Foundation, which financially supported the transformation of science syllabuses in the 1960s, also supported a Unit for the History of Ideas to produce films and textbooks on history of science for sixth form use, the project's Director, Stephen Toulmin, made no links with either the ASE or the BSHS, and the project collapsed in 1964 with only university-level films and texts and a useful *Guide to the History of Ideas* for schoolteachers to its credit.[5]

Recent Developments

That situation began to change in the late 1970s when an effective pressure group of schoolteachers began to express concern that children were not being given the opportunity to evaluate the social and ethical dimensions of science. Two important ASE-supported curriculum projects, the 'Science and Society' project masterminded by the public school science teacher, John Lewis (1981) and the 'Science in its Social Context' (SISCON) project initiated by the comprehensive school teacher, Joan Solomon (1983a), have ensured that in recent years the cultural aspects of science have become an important aspect of 16+ education – not just among youths studying science A-level examinations, but also amongst students studying the humanities as well.[6] To be sure, the historical element in these two courses is small, being largely confined to some history of medicine, evolution, and cosmology; but my point is that they have provided a context for the teacher who is interested in and committed to history of science. More significantly, they have helped persuade the ASE to include the cultural aspects of science teaching in its policy formulations. Meanwhile, in another context, the BSHS has become conscious that the profession of history of science only neglects its non-professional audiences at great peril at a time when government cuts in funding are having a dire effect on history of science in British universities.

Before proceeding to examine the new policy of the ASE and BSHS, we should pause and ask what, if anything, have history teachers been doing about the history of science in schools? In fact, since 1972 there has been quite a revolution in the way history is taught in schools – so much so, that during the last few years fears have been expressed that the pendulum has swung too far against chronology and knowing the 'facts' of history. I refer to the 'Schools Council History 13–16 Project' which began in 1972 as an attempt to break out of the straitjacket of extensive memorizing, which had always been a weakness of school history. Instead, history teachers were asked to place emphasis upon the critical use of evidence and on the evaluation of documentary materials. The fine details of the three-year history course, which ends with GCSE, need not concern us; Joe Scott describes the Schools Council History Project elsewhere in this book. But we should notice that one of the compulsory course elements is a study of the development of medicine from the Stone Age to the mid-twentieth century. This unit is based upon three course booklets and two filmstrips.

Interestingly, evaluation has shown:

1 that the study of history of medicine is one of the most popular segments of the course
2 that the children transferred their learning – but *from* science *to* history, in so far as they made their own links concerning historical and scientific inevitability, cause and functional relationship in science and history.[7]

This Schools Council History programme has now been running in its non-research phase since 1977, which means that many of the students taking science in the sixth form today have learned to ask questions about written and visual evidence and learned to be receptive to the idea that science, like history, is a puzzle-solving activity. Such has been the success of the history of medicine unit that another history of science element has been recently added to the course as an alternative to medicine. Called *Energy through Time* and developed by Joe Scott of the University of Leeds, it was examined for the first time in 1987. It is too soon to say how well it has been received, so I return to the point: history teachers in British schools are teaching history of science in a serious way. In addition, notice that it would be ludicrous now to teach history of science in an old-fashioned formal way, since this method has been repudiated by the 'History 13–16' approach with which the majority of youths are familiar. Hence our history of science must be *problem* and *document* based, and integrated with the science or history teaching in hand.

By 1979, when the ASE published its consultative document, *Alternatives for Science Education*, the implementation of comprehensive education, together with science teachers' experience of teaching and examining for CSE and GCE at 16, had led to widespread agreement that science must become part of the core curriculum of *all* pupils up to the age of 16, and that 'the science curriculum should incorporate a reasonable balance between the specialist and generalist aspects of science' and should reflect science as an intellectual discipline, as a cultural activity, and as a resource which contributes to 'the worlds of work, citizenship, leisure and survival' in today's technological society. In its further policy statement, *Education through Science* (1981), the ASE defined teaching science as a cultural activity, noting:

the more generalized pursuit of scientific knowledge and culture that takes account of the history, philosophy and social implications of scientific activities, and therefore leads to an understanding of the contribution science and technology make to society and the world of ideas.

The policy statement went on to urge schools to try to achieve these aims 'across the curriculum', citing the history lesson as one way of inculcating awareness of science as a cultural activity, while at the same time suggesting how its way of looking at the world differed from other perspectives. As we have seen, the Schools Council's History course had been doing this successfully. Finally, the ASE suggested the development of 'new A-level courses that would appeal to able students who did not wish to specialize in biology, chemistry or physics', and implied that such courses might include the history and philosophy of science.

As far as science curricula generally were concerned, the next and most important recent step was to establish the Secondary Science Curriculum Review under the directorship of Dr R.W. West. This body affirmed in 1983 that in studying science, students should 'gain some understanding of the historical development of scientific principles and theories'.[8] But how was the ASE to proceed with teaching science as culture beyond recommending the SISCON and 'Science and Society' programmes? If history of science was a useful teaching tool, there was the problem that too few science teachers felt or feel competent to handle it. (But note that history teachers had overcome this by being given a comprehensive package of teaching materials to use.)

In May 1980 the BSHS and the ASE agreed that there were three possibilities to consider:

1   teaching the history and philosophy of science as a separate sixth form course;
2   teaching history of science within history courses; and
3   teaching history within science courses.

It was generally agreed that a special sixth form programme of history of science would probably always remain a minority school interest and confined to schools where a teacher held qualifications in the subject and a good deal of enthusiasm for it. Such a person needed little advice. However, whether a $\frac{1}{2}$A-level examination in history and philosophy of science can be devised which is suited to science students while satisfying the national criteria which are now accepted for history courses remains an open question at the time of writing. The second alternative was already happening in view of the 'History 13–16' programme and the changes in history teaching that this approach had encouraged in other syllabuses. How, then, could the BSHS and the ASE help encourage the teaching of history within science?

During 1981 the BSHS held a number of meetings for science teachers, including a symposium at the ASE's annual meeting in Warwick. During 1983 and 1985 the ASE also ran summer schools for English science teachers at the Deutsches Museum in Munich. From this experience it became clear that the best strategy for building upon the interest aroused would be to set up curriculum development groups with the following terms of reference.

1   Groups were to produce and disseminate materials suited to both history and science teachers which would enrich, enliven, and extend the science and history curricula of the 11 to 16 age-group.
2   Groups were to be formed from teachers who taught science *and* either history *or* the social sciences; they would probably already have an interest in history of science.
3   Groups were required to reshape and reorient selected parts of

existing history or science syllabuses so as to improve the treatment of history of science and technology, without, however, adding to the syllabus length.

4   The materials devised could be for pupil discussion or background reading for teachers and pupils, resources in the form of games, case studies, simulation or role-play materials, tape/slide materials, or materials for the in-service training of teachers, or simply articles for publication in teachers' periodicals.

5   Finally, the curriculum groups were specifically directed to attend to the history of local industry and technology.

Illustrations

By way of example let me mention briefly the teaching modules which Welsh teachers have produced on the chemical industries of South Wales. These 20-hour teaching modules aim to use history to stimulate interest in science and technology among 11–16-year-old children. Their aims and objectives are:

1   to broaden the scope of the teaching of science, particularly chemistry, to embrace local industrial history;
2   to consider the relevance of science and technology to the way in which the local industry developed;
3   to make students aware of contemporary man's dependence upon the chemical industry; and more generally, of how technological developments also change the lives of communities;
4   recognition of the remains of past industry, and hence to make pupils more aware of their environment.

By combining lessons with laboratory and field work, the courses also aim to help students collate, present, and interpret evidence and enhance observational skills, while at the same time providing them with laboratory and fieldwork experience.

The modules are packaged as combined teacher–pupil loose-leaf booklets and accompanied by worksheets and overhead and slide materials. They clearly demonstrate what can be done at a local level to use material in the history of science and technology.

Similar courses on the iron and steel industry have also been developed in certain London schools by Sheila Martin. Here the intention has also been to introduce city children in the privileged south of England to the nature of chemical industry, and the courses have actually involved taking London children to South Wales on a field trip. In another London school the interrelated themes of the nineteenth-century synthetic dyestuffs industry and the story of nitrogen fixation in this century have been used in general science and chemistry teaching to enhance industrial awareness as well as scientific learning among children in the 11 to 14 age-group. Two teachers' monographs and pupils' packs, *The Colour Chemists* (1983) and *High Pressure Chemists* (1984) have been issued by the ASE. The ASE has also recently published a module which deals with the history of pollution and its control, *Topics on Pollution*, for sixth form pupils.

From what I have said it is clear that most of the experimentation with history of science in the science curriculum has actually involved history of technology, industrial archaeology (and perhaps history of medicine, rather than the history of scientific ideas). Indeed, although I have seen suggestions for modules on the inert gases, and on the teaching of organic chemistry, these suggestions have yet to be developed effectively by teachers themselves. It is also the case that although the South Wales modules I have described have been developed for an all ability range, their primary target has been the ordinary, non-academically motivated child in the comprehensive school. In other words, we are not aiming to teach history of science per se, but to use it as a tool for awakening interest and love of science, to the benefit of the scientific literacy and decision making process of future generations

Conclusion

Let me end by mentioning briefly the very latest developments in Britain and by suggesting what the future may hold.

Both the Royal Society of Chemistry and the Institute of Physics have played an active role in stimulating historical interests among their memberships. They have done this through their publications *Education in Chemistry* and *Education in Physics*, and through their respective Historical Groups. The RSC History Group, for example, has produced an elaborate bibliography, though, arguably, this is of greater use to the professional historian of chemistry than to the schoolteacher![9]

Other recent developments have been the publication in 1987 of a

Teachers' *Handbook* on teaching the history of science[10], and the BSHS's decision in the same year to set up a special membership category for school science teachers. The Open University's excellent film library of history of science and technology will, it is hoped, be made available, in the form of specially compiled and edited low-cost videos.

What, then, of the future? To put this new enthusiasm for the history of science to work, teachers will need informed, interesting, and challenging materials which they can use with pupils. Such materials, as shown by the examples I have given, will need to be complete within themselves and adaptable to classroom conditions. British teachers have little time for curriculum development. They therefore need complete digests of information: here is the information; these are the facts; this is the story; this is how you can use the material.

In conclusion, the future looks bright for the increased use of history of science in British science teaching – especially now that the government and the ASE agree that a wider range of approaches to the teaching and learning of science is required. Indeed, the Minister of Education's determination to introduce a common core curriculum into British secondary education may well act strongly in favour of history, philosophy, and social studies of science because, for the first time, pupils will have to continue with science lessons until the age of sixteen. And since 'academic science' will suit only a minority of pupils, 'science for all' will be forced to concentrate on the *processes* of science – upon how scientists reason and think, how they reach conclusions, and on how their methods and techniques are affected by society and affect it. And, surely, these processes will be best studied from case histories drawn from the recent, as well as the distant past, and not from some idealistic notions of how scientists work.

The new criteria for the continuous assessment of pupils for GCSE examinations may also stimulate teachers to situate the practical work which has to be repeated for assessment within a historical framework. It seems, then, that history of science will at last have found its place as an adjunct to science teaching, both to act as a midwife to learning and comprehension of concepts and techniques, and to provide a deeper understanding of science in its social and cultural context.

NOTES

1 A shorter version of this chapter has been published in German. See Brock (1988), pp. 57–71. For the early part of this chapter, I am much indebted to

the thesis of my former student, W.J. Sherratt (1980, 1982–3). Sherratt's attitude survey of a sample of schoolchildren, although unsatisfactory, did suggest that some children welcomed, and benefited from, history of science.
2   Mansell (1976).
3   *School Science Review* (1935–6).
4   These were O/A-level examinations designed to be taken by science or arts pupils at the age of 17 at the end of their first year in the sixth form. The papers, which were set by the Oxford and Cambridge Joint Examinations Board, and by the Cambridge Local Examination Board, were amalgamated in 1982. Plans to transform the examination in 1988–9 into an 'Advanced-Supplementary' paper to be set jointly by the same two Boards together with the Southern Examinations Board were rejected by the Secondary Examinations Council as not meeting the national criteria laid down for history examinations. The revised syllabus had been criticized at a three-day meeting of science teachers in Cambridge in 1986 (see *Education in Science* (1987), January, p. 28, and *History of Science* (1963), 84–90). Consequently there is now no examination at this level in history of science.
5   Steele (1970).
6   There had been an earlier Schools Council 'Science and Society' project (1972) which used portfolios of newspaper and journal articles to stimulate Sixth Form discussion. But none of this material was historical. See *Science and Society* plus a teacher's guide by John W. Lewis (1981); and *Science in Social Context* plus a teacher's guide by Joan Solomon and Sue Addinell (1983).
7   Shemilt (1977, and 1980), Scott (1986).
8   Secondary Science Curriculum Review (1983).
9   Russell (1985).
10   ASE (1987). This was the final product of the Joint Working Party of the ASE and BSHS established in May 1980 to stimulate the use of history of science in schools.

Teaching the History of Science
Is Nothing Sacred?

Joan Solomon

We all love the history of science but this should not blind us to the possibility that, like olives or dry sherry, it may not be equally well loved by the young. My own experiences, as an enthusiast trying to teach the history of science to the average ability comprehensive school pupil, suggest that it may indeed be an acquired taste. This implies that we should consider with the greatest care:

Why do we want to teach it?

How are we going to go about it?

These are difficult questions. The first is hard for the same reason that the psychiatrist's chair is uncomfortable. I suspect that we would all like to cover our real personal motives with rhetoric about 'culture' and 'insight', but a few periods with the mixed ability Thirds on a wet Thursday afternoon would soon show up the holes in that garment. Even if we have more enduring and iron-clad aims and objectives the Thirds still wait for their lessons and we shall need to address the second question with a good deal of industry. This article will attempt to provide some answers with help and evidence from a wide variety of sources both within and without science.

Why Teach the History of Science?

Just propaganda?

We shall need to answer this question in a very hard-nosed way in the context of the school curriculum which is a 'zero sum' affair. Every new subject that we introduce must oust another, just as Technology and Microelectronics are now in the process of doing. If we try to tuck it into corners of the main science curriculum some other content or process will have to go, so we must be well prepared for counter-sniping. Arguments that it is a complete discipline on its own, offering a whole range of valuable life skills that other subjects also offer, will not do. We need a stronger, special purpose.

If that sounds suspiciously like using the history of science as propaganda, then we should admit to both the possibility and its long and venerable history. For centuries it was used to glorify science itself; Simon Schaffer quotes Priestley's presentation of this study as 'the human understanding grasping at the noblest of objects' while social history merely reports 'what is most tedious and disgusting'.[1] Fortunately such wholesale chauvinism is not entirely necessary. It has also been suggested that the recent history of science and technology should be employed to inform and influence science policy makers,[2] and to convince teachers of the value of balanced science within schools.[3] For what would school children need it?

As culture?

Ever since C.P. Snow's exposure of *The Two Cultures*, science has taken to the battle for cultural respectability on its own terms. This argues for teaching not just literature *and* science but the literature *of* science, not history *and* science, but the history *of* science. My own teaching and writing on the history and philosophy of science began in this atmosphere of scientific challenge to establishment culture, as did that of many others. In the USA a similar interest was being shown at about the same time. Klopfer developed case study materials for school use in the late 1950s, and his argument for 'scientific literacy' was crystallized in a seminal paper which argued for understanding 'the interaction between science and general culture'.[4] The same period saw the publication of the *Harvard Physics Project* which gave history of science a greater prominence than any other science course, before or since.[5]

Today this belief in science as culture is buoyant enough to surface in almost every policy document on science education (although probably is much rarer in fact than in polemic). The influential ASE document *Education through Science* (1981) advised teaching

Science as a cultural activity: the more generalized pursuit of scientific knowledge and culture that takes account of the history, philosophy and social implications of scientific activities, and therefore leads to an understanding of the contribution science and technology make to society and the world of ideas.[6]

This has been widely quoted,[7] although it is not entirely clear whether the history of science is being justified for its understanding of contributions made to society, or to the world of ideas. It is the latter claim which is usually to be found under the cultural banner and which is, sadly, quite the hardest to justify to our typical recalcitrant Third former, for whom the word 'culture' is more usually prefaced by either 'youth' or 'pop'.

For illustrating the nature of scientific knowledge?

During this century the history of science has taken priority over its philosophy. Kuhn and Popper, as philosophers of science, have buttressed what they have taught us about the nature of scientific evidence not by allusions to logic or the nature of truth, as would have been done in previous centuries, but by examining how actual theories have risen and fallen in the estimation of the scientific community. Much the same could be said about the sociology of science or the new iconoclastic studies of non-science and parascience. If the history of science is thus the accepted source of understanding about the scientific method, does this not provide a clinching argument for its use in school? Certainly both Klopfer, for school students, and Conant[8] for university students, made precisely this claim.

But to rest the case on Kuhn's insights and philosophy brings its own traps. Either we teach about normal science, which is worthy but may lack drama, or we teach about some important scientific revolution and come up against incommensurability. From this side of a Kuhnian paradigm shift what went before is always hard to understand and liable to be dismissed as trivial, wrong, or both. Anyone who has ever tried to teach about the phlogiston theory to the modern Fifth former (let alone the terrible Thirds) will know just how great an effort is required to show the power of this bygone explanation to pupils who may have just achieved an A grade for getting the oxygen theory *right* in the last end-of-unit test.

One reason why it is so hard to teach about a scientific revolution is not only the paradigm shift, but the placing of this learning-and-unlearning so near in time to the pupil's own uncertain attempts at learning the very same content. Two suggestions have been made for the treatment of this problem. One is to reorganize the school science curriculum so that it retraces the history of science. This is recommended on the grounds that it would escape from the 'learning and unlearning' bind, but reflection shows how very muddling and time-consuming such a syllabus might be. Even optimistic curriculum evaluation has cast doubt on the effectiveness of this approach.[9]

Others have tried to use evidence from current research into the common misconceptions of school students in science lessons to suggest that their own notions are so like the displaced theories of earlier times that the arguments of a past scientific revolution might be just what is needed to effect a useful cognitive change. But closer examination shows the fallacy of this. Children's ideas are not complete theories consistently applied; bits of their explanations may sound like Aristotelian mechanics, other bits like medieval impetus theory, and yet others may be idiosyncratic and animistic.[10] Let us not pervert history by comparing the past theories of natural philosophers to the early untutored attempts of a modern child. (Galileo did something very like this in the full-blooded polemics of *Dialogues concerning Two World Systems*, but that is no excuse for us to follow his example.)

Nevertheless, even if the going will be difficult, this argument that we need to teach about the nature of scientific knowledge, and that they way to do it is by stories from history, may stand up so strongly that it must be accepted. As the pendulum swings in educational fashions we have been carried, in recent times, away from *concepts* and *theories* and towards the *processes* extreme. It has been widely suggested, by those who should know better, that a continual emphasis on experimental work and on its assessment for examination purposes is all that is needed to teach about how science works. Those few researchers who have explored what school students think about science in the context of important social issues have indicated that this naive positivist school-learnt attitude serves our students very ill.[11] When difficult issues arise and scientists do not give us the 'right answer', it must show, such students are forced to suppose, either that scientists are charlatans, or that they have not done the right experiments, or that they are keeping 'the truth' from us. Simple extracts from the history of science may be just what is needed to start putting that right.[12]

To show the human face of science?

This argument for including some history of science in our school science courses is based on motivation rather than new learning. The comprehensive review paper by Thomas Russell[13] makes it clear that while the verdict of research may indicate a higher score on affective values such as 'course satisfaction' and 'book enjoyable to read' when the historical aspects are emphasized (see, for example Welch and Warburg's study of students on the Harvard *Project Physics* course[14]) no associated gain in the understanding science is likely to be recorded. Russell echoes much of the general educational disquiet about results of attitude tests but nevertheless concludes with recommendations that more history of science should be included.

One of the major purposes that Russell sees in teaching through history is to correct the image that students have of the scientist himself (the use of the masculine form here is quite intentional). Almost every study has shown that this image is anti-social, eccentric, irresponsible, and strongly influenced by comic strips and TV cartoons. But teaching to correct such a 'pop' image is neither easy nor entirely in the current stream of scholarship. Modern historians of science repudiate the image of 'scientist as hero' which figured in the inspiring books of my own childhood within the general genre of 'From the Greeks to John Dalton'. When John Ziman addressed the question 'who was a scientist?' in his excellent book *The Force of Knowledge* he had to answer simply 'They were very *individual* people'.[15] In our school science lessons we shall not be able to do much better than recount diverse stories of diverse people. Will that be enough for the purposes of motivation?

In some ways this simple emphasis on the people who carried out the scientific enterprise may be just what we need in school, even if it runs counter to modern emphasis on the history of institutions and instrumentation. The recent effort to attract girls into the physical sciences has given new incentive to the study of attitudes. What the careful research of Collings and Smithers, and of Pell has shown is that it is 'person orientation' which most clearly distinguishes the majority of girls' interests from those of the boys', and also differentiates sharply between the non-scientists and those who have chosen to study science.[16] There is also some indication from the work of John Head on adolescent boys that this gender difference may be most clearly defined at the very age when the school subject choices have traditionally been made. Now that the Department of Education and Science has recommended 'Balanced Science' courses in which all pupils encounter all sciences up to the level of GCSE, we have a new need to change our teaching.[17] Choices will not be made: both girls and boys will have to study the physical sciences. Unless we choose to ignore the preferences of at least half our classes we shall need to show the human face of science and scientists in our lessons.

Of course scientists will not be heroes to our pupils the way in which Superman and gun-handling car-toppling cops may be, and it would be an appalling mistake to present them as such. I am reminded of a wet afternoon lesson some ten years ago when, having to substitute at short notice for an absent history teacher, I unwisely set out to tell the story of Galileo's trial to a mixed ability Third Year class. They listened with gratifying attention to the story up to the moment when Galileo – old, blind, and kneeling on the cold stone floor of the monastery – was shown the instruments of torture. Then when I told how he publically recanted his heretical views there was a rustle of discontent. My pupils lost no time in making clear their disgust with such cowardly anti-heroic behaviour!

Scientists are heroes because of their discoveries, not their bravery, honesty, or clean living. Count Rumford was an exhibitionist and a traitor but his story of boring the canon is in the heroic tradition. Humphry Davy may have faked the ice-rubbing experiment but the image of him dancing round the laboratory in jubilation when he first separated potassium shows the 'glory' of heroic discovery in a perfectly human way. Anyone who saw the recent TV film 'Life Story' will remember the similar antics of Watson and Crick when they solved the enigma of DNA as far more inspiring for young or old than any walk alone at high noon. The personal touches that we need from the history of science are in the human context of real and heart-warming discovery, the kind that many of us got in our youth from *The Microbe Hunters* or *Microbes and Men*.

For understanding the social relations of science?

It could be argued against the case made for motivation that spectacular scientific discovery is not likely to be the lot of most of our pupils. Perhaps it is even some kind of 'con' to use bait which hooks fish to a purpose which they cannot serve. It would be more satisfying if we could find a reason for teaching the history of science that had an importance beyond just happy motivation.

We are fortunate, as science teachers, in not being forced to argue desperately for our discipline in the present job-hunting educational marketplace. It is far worse for history teachers. Most of them have to make do with the rather thin 'cultural' argument that was dismissed some three sections back. That makes it all the more impressive to read Dawson's recent fighting defence of the School History Project: 'The Project's philosophy stands or falls according to its success in demonstrating to pupils that historical understandings are transferable and can enhance their lives outside the classroom . . . [by] applying historical thinking to pupils' lives in the 1980s'.[18] This, it seems to me, is the hub of the challenge. Are there lessons from the history of science that we can honestly recommend to all our pupils as of value in their daily life as citizens rather than as the scientists that most of them will never become?

The public's understanding of science, and its participation in science-related issues, has recently become of sufficient importance to figure in a report of the Royal Society, and to be the subject of a number of related pieces of research. In school such issues have figured prominently in courses like *SISCON in schools* (1983) which I have taught for several years. In these there is a special and valuable place for the history of science: each of the booklets contains substantial amounts of such material at the start of each topic. The purpose of this is to show the complex nature of the interaction between our science, our technology, and our society in a way infinitely more acceptable than any dry lecture on this theme.

Suppose, for example, that you want to teach about technological innovation in such a way as to bring out the role of market pull, patents, scientific knowledge, scaling-up problems of production, industrial safety, the fears of society, military pressures, and the responsibility of scientists. All of these are hard to grasp and define, but telling a story like that of the 'smokeless explosive' from the *Technology, Invention and Industry* SISCON booklet[19] both holds the attention, and also allows the pupils themselves to bring out the ideas that we want.

The point is that incidents from the past allow sufficient perspective for us to reflect upon social and technological factors; present crises are sometimes too charged with feeling for us to see them clearly. In a similar spirit the SISCON booklet on nuclear armaments traces the history of the development of the atomic bomb, the one on space technology describes early science fiction, rockets, and interest in interplanetary flight. The booklet on public health recounts stories of nineteenth-century pollution in the Thames, the thalidomide disaster, and the development of the contraceptive pill. From a different stable the splendid case-study on the development of coal mining in north-east Britain, *King Coal*, describes through contemporary photographs and documents how appalling working conditions and industrial pollution arose and were gradually changed by social pressure.[20]

As an introduction to social history such materials are clearly excellent in their own right, but as sources from which we can begin to demonstrate the benefits and risks of the social interactions of science and technology they are invaluable. This is the area in which it is easiest to demonstrate the importance of the citizen's understanding of science. Unfortunately it is also one in which the underlying ideas, which might be transferable to those unknown future events in which our pupils will have to decide and act, are quite the most difficult to expose. We can teach science in the laboratory, and technology in the workshop, but the social relations of science only begin to come to life in a comprehensible way through a careful selection of case-studies from the past with which our pupils can compare the problems of the present.

How Shall We Teach It?

Lessons from the teaching of science

The most unfashionable thing to do in science lessons is to lecture the pupils. Every educational authority has been busy telling us not to do this. There are twin reasons for this reproach: one is the boredom of the pupils and the other is their lack of that active involvement by which they may 'construct' the knowledge. In science there may be a tendency to think of this construction of knowledge solely in terms of experimental work, but at heart this argument is about ownership of knowledge. However it comes about, the pupils must have contributed enough of their own thought, feeling, talk, and action for them to have personalized what has been offered.

This is a severe constraint on teaching the history of science. We may not lecture and may only spin stories for short lengths of time. Some have argued that we could use the old science lesson format by arranging for the class to carry out the crucial experiment for themselves. Certainly there are times when this can be done, like dropping different masses (even if this was not performed from the Leaning Tower of Pisa by Galileo himself, it did happen at about this time), and it is usually as surprising and thought provoking today as it was in the seventeenth century. But many famous historical experiments are either difficult to perform like Michelson and Morley's, dangerous like Franklin's, Priestley's and Lavoisier's, or dependent on newly made but now old fashioned equipment, like Volta's pile. Replace the apparatus with what pupils would expect to find in the school laboratory and the element of history begins to become very tenuous.

Nor are outings to museums to be equated with scientific field-trips. Traditional museums do not encourage pupils to take over the ownership of ideas. (Glass cases of labelled 'eoliths' are irremediably boring if they may not be touched or, better still, be used as a challenge to make similar collections.)

Skills in the learning of history

There is no need to re-invent the wheel. Our colleagues in history education have not only been struggling on a daily basis with the mixed ability Thirds, they have also been putting their conclusions into print. In his inspiring essay on *Imagination in Teaching History* John Duckworth started by setting aside the scientific and analytical approaches to history. This, he suggested 'leaves two courses open to the history teacher and his pupils; firstly an individual appraisal of the 'evidence' and secondly the seeking of imaginative insights into the past . . . How do we excite an imaginative response in history?'[21] By 1984 Culpin reported that although the theoretical reasons for developing the twin skills of imaginative empathy and interpretation of evidence in the 'New History' teaching were well established, they were still far from common in the classroom. Primary and secondary source materials were often presented but engendering historically valuable empathy was more difficult. Indeed the latter skill is itself quite vital for the interpretation of historical material: 'A piece of empathic work is certainly a creative response . . . But it is also a historical response . . . It is not an act of untrammelled fantasy; it is, at best, a meticulous piece of historical work.'[22] Dawson has analysed this quality of historical empathy into three parts: the roles of individuals (both for causation and for social complexity), contrasts and similarities between then and now, and analyses of sources within their context. The first of these may be of the greatest importance both for showing the heroic nature of personal scientific discovery, and also for illustrating the action of science and technology on social conditions.

It is no problem for history and STS teaching that didactic conclusions about the rightness or wrongness of particular incidents or conditions cannot be made. There is immense educational value in having children argue about the social justice of the Factory Acts or the ethics of Jenner's original vaccination experiments. These are exactly the kinds of activity which engender empathy and involvement just because they are open-ended questions with no right answer. (In the same way STS students are encouraged to take sides in arguments about testing for AIDS and the generation of electricity from different primary fuels.) What history and STS teachers are both trying to do is to get '[their] pupils to put themselves into situations of conflict where absolute standards of right and wrong cease to have meaning'.[23]

The requirements of the national criteria for GCSE are that all science courses shall include considerations of the social implications of science. For teachers of the history of science this will slot very nicely into the methods they use for both the historical and the contemporary aspects of their lessons. For the students it involves considerations of social justice and citizen responsibility, as well as of scientific and historical content, and so should greatly enrich their learning.

Methods of teaching

There is little doubt that journeymen teachers of the history of science will find it difficult to reach the ideals set out above but they do give some clear indications of the ways in which we should be going. Aims and objectives are only the starting point for teaching. In what follows the pros and cons of four different methods will be very briefly considered:

1   using contemporary sources,
2   using drama,
3   small group discussion work,
4   imaginative writing and art-work.

The diaries of scientists, or their writings, make fascinating reading, but often they are too demanding for our students (although, perhaps, a glance at the transcript of Faraday's laboratory book might help convert those infuriating 'I will copy-up later in neat' pupils). Photographs and pictures are far better, but pupils need a clearly defined task, either written or oral, if the act of looking is to move from museum-type observation towards empathic activity.

Drama springs to mind as soon as empathy and involvement are mentioned, but it is a far more difficult classroom activity to manage than most assume. There must be no fighting because the sham so often turns to real, there must be almost no learning of lines because that takes away individual acts of involvement, and there must be no more than short bursts of watching the halting and amateur efforts of others. We are left with either short pieces of mime, or short invented dialogue. Embedding such episodes into an enjoyable and dramatic whole suggests witnesses in a trial, or eye-witness reportage after an incident.

Giving speeches comes hard to most school-age children, and doubly so if it takes place in the classroom in front of their peers. Nevertheless oral discussion work is an important and much neglected educational skill. Most of our pupils will hardly ever write again after they leave school, but they will talk and argue. If we could give them at least a little group practice in listening to the arguments of others, answering points, showing where they disagree, and advancing alternative arguments of their own, then they would have acquired something that could indeed be called, in the modern jargon, a 'life skill'. In STS and social history courses, where evidence is open to different interpretations and to different ethical verdicts, small group discussion with a recorder and a chairperson is a well-tried and successful method.

Imaginative writing and appropriate art-work are far from new in the history classroom, although still unusual in science. Because writing, and to a lesser extent drawing, is a private activity in which friends and neighbours do not take much part they do little to break down the walls of isolation within the classroom. For a few of our students, who are shy or not well-liked, such classwork may be valuable: more often these activities are suitable for carrying out at home. But if historical materials are present to help develop the necessary context they can be shared and talked over between friends, thus adding to the pleasure and value of the work.

Conclusions

This chapter has looked at the teaching of history within the mainstream science syllabus, not as a minority activity for a single subject. That is the hardest case to make. To further it we would have to face our colleagues with suggestions of what science to cast out to clear space for it. Yet common sense insists that without some understanding of science its history is valueless. If neither the arguments for 'science as culture' nor for 'scientific method through its history' are tenable as education for our mixed ability Thirds then all that is left is the case for 'citizen science' which uses history to understand the contemporary social issues of science. But perhaps, after all, that is no bad argument.

NOTES

1 Schaffer (1985).
2 Pickstone (1985).
3 Solomon (1987).
4 Klopfer (1969), pp. 87–95.
5 Rutherford, Holton, and Watson (1970).
6 Association for Science Education (1981).
7 Moore (1987), SSCR (1987).
8 Conant (1951).
9 Klopfer (1969), pp. 87–95.
10 Champayne et al. (1981), Solomon (1983a).
11 Fleming (1986).
12 See, for example, the SISCON in Schools Project booklet 'How can we be sure?' (Solomon 1983b).
13 Russell (1981).
14 Welch and Warburg (1972).
15 Ziman (1976).
16 Pell (1985).
17 Department of Education and Science (1985).
18 Dawson (1987).
19 The SISCON in Schools Project booklet 'Technology, Invention and Industry' (Solomon 1983b).
20 Gateshead Department of Education (1982).
21 Duckworth (1971).
22 Culpin (1984).
23 Duckworth (1971).

History of Science and Science Education

Stephen Brush

How can we improve the general understanding of science? The current revival of concern for the quality and effectiveness of education includes discussion of issues such as motivating students to learn subjects they regard as difficult, changing the public perception of scientists, encouraging informed participation in decisions about the uses of technology, and conveying an appreciation of science as part of culture. To achieve these goals, educators have often recommended the use of the history of science.[1]

In this paper I review briefly what history of science has to offer science education, and in particular how it has already been used in a successful science curriculum project.

First, let me remind you that 'history of science' is not just a collection of books and articles waiting to be pulled off the shelf and plugged into the curriculum. Like science or education, it is an ongoing enterprise carried on by people with their own goals that may or may not include improving science education.

Over the last 50 years, history of science has evolved from a subject studied seriously by only a few scholars but widely used in science teaching to an established academic discipline somewhat isolated from the scientific community. Professional historians of science, seeing themselves as historians rather than scientists, criticized scientists for promulgating 'Whiggism',[2] and some of them overemphasized the social context at the expense of the technical content of science.

Now, having attained a degree of maturity and acceptance within the humanities and social sciences, history of science has started to rebuild the bridges to science. Historians of science welcome scientists with an interest in history and offer a variety of materials for explaining science to students and the public. Scientists for their part have a better understanding of the value of sound historical research, and have given substantial support to historical sections and centers in their own societies, as well as reading scholarly books and journals on topics in the history of science. This is especially true of physics, but other sciences have also moved in this direction.[3] The public-relations as well as the pedagogical benefits of history are now recognized.

Misunderstandings still hinder communication between different academic disciplines. It is disturbing to find that some prominent educators use the label 'history of science' in an obsolete and misleading sense to characterize the *traditional* approach to science teaching, and blame it for the failings of that approach. In the latest edition of a widely used reference work on educational research we find the following passages in a summary of reports on science education:

In the 'historical-survey' approach, the teacher is responsible for exposing students to a review of the conclusions scientists have reached in past years. Faced with few resources, little time, and a textbook filled with glimpses of past scientific accomplishment, the teacher copes with a 'science history' in primarily an assign–recite–test mode of teaching . . .although student motivation is very low, this is excused, because science is supposed to be a rigorous and hard subject . . . At the middle–junior high level, the studies showed, science was taught as a survey of science disciplines – that is, a 'history-of-science-concept' course . . . the science curriculum was essentially a 'read-and-repeat' of a science textbook mixed with infrequent demonstrations of phenomena . . . Strong control and discipline were much easier to obtain and maintain if students were reading and repeating than if they were actually involved in experimenting and thinking.[4]

This is exactly the *opposite* of what I mean by a historical approach! The essence of such an approach is not merely to assert the conclusions but to show how they were reached and what alternatives were plausibly advocated; the historical approach certainly demands thinking and, in many cases, experiments.

Twenty or thirty years ago, in a discussion of this subject, one would have described the benefits that *might* be gained by using an historical approach to teach science, with only anecdotal evidence to support one's claims. But today one can point to a major curriculum project, the Project Physics Course developed at Harvard with support from the National Science Foundation, as a successful example. The text does not simply present the conclusions of research but shows how they were obtained; the course does not rely entirely on the text but has a full array of experiments (including simulations of historical experiments), demonstrations, films, transparencies, and other resources. Several of the people involved in writing the Project Physics text (including myself) were historians of science, as indicated by their publications and current academic appointments, though they also had advanced training and research experience in science.

According to its Directors, the three major goals of Harvard Project Physics were 'to design a humanistically oriented physics course, to attract more students to the study of introductory physics, and to find out more about the factors that influence the learning of science.' Of the several effects they hoped to achieve, the first two were

1. To help students increase their knowledge of the physical world by concentrating on ideas that characterize physics as a science at its best, rather than concentrating on isolated bits of information.
2. To help students see physics as the wonderfully many-sided human activity that it really is. This meant presenting the subject in historical and cultural perspective, and showing that the ideas of physics have a tradition as well as ways of evolutionary adaptation and change.[5]

After it became commercially available in 1970, the Project Physics Course was widely adopted throughout the United States. Quantitative estimates of the number of schools and students using various physics texts are not very consistent with each other, but a reasonable guess is that in the mid-1970s about 15% of students taking physics courses were using the Project Physics Course.[6] This is a substantial achievement in the view of the widely lamented fact that much of the federal funds poured into curriculum development in the 1960s and early 1970s led to products that reached less than 5% of their intended audience.[7] The introduction of the Project Physics Course did not accomplish the goal of increasing the *total* enrollment in physics courses throughout the country; its gains seem to have been at the expense of PSSC (the course developed earlier by the Physical Science Study Committee) and other courses. There is some evidence that it did increase the enrollment in those schools where it was used,[8] so that it may have helped to offset the general decline in enrollment elsewhere. The course has been translated into other languages and continues to be used in other countries, such as Italy.

The staff of Harvard Project Physics included several specialists in educational research and evaluation, and there was from the beginning a substantial effort to determine whether the course was accomplishing its objectives. After 1970 the use of the course in many schools attracted research projects by other investigators not affiliated with the original project. Some of these studies looked specifically at goals that the historical approach was intended to achieve.

The earliest studies showed that the Project Physics course significantly changed students' *attitudes* to physics: they found it more historical and philosophical than those who took PSSC; they concluded that the course itself was less difficult and mathematical, and the text was more enjoyable to read.[9]

Glen S. Aikenhead found that 'in four areas specifically (the tactics of science, its values, its institutional functions, and the interaction of science with society), the HPP achievement *greatly* exceeded that of the control group'; students who took the Project Physics course learned not to think in terms of a standardized 'scientific method' but gained an appreciation of the roles of diverse approaches, imagination, confirmation, and instrumentation in the pursuit of scientific knowledge.[10]

David J. Quattropani studied the change in student understanding of the relationships among science, technology, and society in eight Project Physics and ten PSSC classes in Connecticut. He reported that the Project Physics group did improve significantly in this respect, although the degree of increase found in this group was 'not enough to establish a significant difference between the two groups.' He concluded that 'an historical approach to science holds promise and should be explored further as a method of increasing student understanding of the relationships among science, technology and society.'[11]

Are these gains in understanding the nature of science and its social aspects purchased at the cost of a decrease in the amount of science *content* learned? Does injecting history merely serve to 'water down' the course and divert time and energy away from 'real science' towards 'frills'? Some critics have indeed pointed out that the Project Physics Text fails to cover all the topics in the traditional textbooks, even though the same critics call its treatment of basic scientific concepts 'excellent.'[12]

Insofar as the acquisition of scientific knowledge can be measured by the College Board Physics Achievement Test, the answer seems to be that students who take the Project Physics Course do nearly as well as those who take other courses. Analysis of the test results for the period May 1970 through January 1974 showed that Project Physics students had average scores ranging from 570 to 608 (the maximum is 800), while those taking traditional physics courses had averages from 583 to 609.[13] Of course the latter is not a true 'control group' since students with low interest or ability in science may be advised to take Project Physics whereas bright students with a strong interest in physics might be advised to take traditional courses or PSSC. Thus one would expect that Project Physics students would score significantly lower even if the text were as effective in conveying technical information. The reason they don't score significantly lower is revealed by another experiment. The New York State Regents Examination in Physics was given to a national sample of students, and it was found that Project Physics students in states other than New York did not do as well as other students, but within New York State they did just as well. New York teachers, knowing that their students would be facing this exam which would affect their academic future, simply took a few class periods to drill them on subjects they knew would be on the exam but were not covered in the text.[14] Presumably the same tactic works for students who are going to take the College Board exam.

Of course one needs more than a textbook, no matter how effective, to improve learning. According to Wayne Welch, most research studies indicate that no more than 5% of the variance in student learning is due to the curriculum. In using the historical approach, the role of the teacher is crucial. Yet most science teachers have little or no training in the history of science. The abolition of summer institutes, one possible means of providing this training, was a severe blow to the long-term success of the Project Physics Course. This or some other means of providing a background in history of science for science teachers seems essential if we are to realize the full benefits of the resources already put into developing the course. But there is nothing wrong with a piecemeal approach, treating some topics historically and the rest in a more traditional way. Given limited time and opportunity most teachers might better aim for an enthusiastic well-informed presentation of one episode rather than a superficial chronology of the entire subject.

On the college level many science teachers use the historical approach, especially in courses for non-majors. A number of books and articles on the history of science are available for such courses. In addition, many colleges have one or more professional historians of science on their faculty available to advise science teachers as well as to give their own courses.[15] But these teachers also need more opportunities to enhance their own knowledge of the history of science. The large attendance at historical sessions held during American Physical Society meetings, and the number of applications for places in occasional NEH Summer Seminars on history of science, indicate a substantial demand for faculty development in this area.

The Education Committee of the History of Science Society has for several years sponsored symposia on the use of history of science in college curricula.[16] The most recent result of this effort is a guide for teachers, providing suggested readings and bibliographies as well as synopses of several important topics in the history of modern science.[17]

I have not said much about the specific features or example of the historical approach that might be useful in teaching science because I and others have written extensively on this subject in the past,[18] but it is appropriate to mention some aspects relevant to science education in the late 1980s. As Thomas L. Russell points out in a recent article, the experience of the Project Physics course shows that we *can* improve attitudes toward science and the understanding of how science works by the use of history. But if we want to improve students' understanding of science itself we must treat the historical material 'in ways which illuminate particular characteristics of science.'[19] Here are three such characteristics I would like to stress.

1   Science deals with broad philosophical questions that should be of general interest even though they are often ignored in textbooks. These questions are best discussed in their historical settings. Introducing them in the classroom may help to counter the ubiquitous tendency to judge science primarily on the basis of its practical applications (a tendency encouraged, somewhat foolishly, by scientists themselves).

In the eighteenth century a major issue was whether the laws of nature, in particular those of Isaac Newton, are sufficient to account for the phenomena of the physical world such as the motions of the Moon and planets – the 'clockwork universe' theory – or whether some kind of divine intervention must be invoked to prevent the system from collapsing (as Newton himself apparently thought). This issue was closely related to the religious and philosophical debates of the French Englightenment, and to some of the political ideas that went into the drafting of the US Constitution.

In the nineteenth century one of the most important questions was whether the Earth and its inhabitants were created only a few thousand years ago or have evolved in accordance with natural laws over periods of hundreds of millions of years. That question involved not only the foundations of biology and geology, but also the development of history and the social sciences as academic disciplines.

In the twentieth century, physicists argued not only about whether the behavior of subatomic particles is inherently random, but also whether it is even meaningful to say such behavior *exists independently of our observation of it*. Mathematicians disputed the possibility of proving anything with complete certainty.

To bring up such issues for discussion does not by any means imply that one has to give 'equal time' to both sides. On the contrary, what one learns from history is that one side was victorious for reasons that can be specified, and thus earned the privilege of having its position adopted, not as immutable truth but as the best available working hypothesis on which to base further research. We want to make the point that scientists are often most successful when they act as if some law, theory or even philosophical principle had been firmly established, even though they recognize that it may be drastically revised sometime in the future.

2   Scientific research combines the discovery of objective facts about the world and the creation of new concepts to describe and predict those facts. Traditional science teaching and popular journalism give too much emphasis to the former, thus perpetuating the image of the arrogant expert who has access to infallible knowledge about the world ('science has shown that . . .'). It goes along with the claim that all the phenomena of nature including human behavior can be reduced to atomic particles and physical forces. This position is often labelled 'positivism,' 'scientism,' or 'mechanistic reductionism' by critics of science, and is sometimes blamed for negative attitudes of students and public.[20]

At the other extreme, some sociologists and philosophers claim that all knowledge, including scientific facts and mathematical theorems, is 'socially constructed.' From this veiwpoint science has no claim to a preferred status over astrology, religion, mysticism, creationism, and other doctrines.[21]

An important task of science education is to strike a balance between these two extremes – to show how science can acquire valid and useful knowledge that is nevertheless a product of human thought, subject to change in the light of new evidence and reasoning.[22]

Two contrasting historical examples illustrate the problem: the Copernican system and the IQ test. According to modern historians of astronomy, Copernicus adopted a heliocentric system not because it was quantitatively more accurate than the geocentric system available at the time, or because it was substantially simpler. (The familiar idea that the geocentric theorists had to keep adding 'epicycles on epicycles' to fit new data is completely mythical.) Instead his reason was primarily philosophical, even esthetic – he wanted to go back to the old idea of uniform circular motion.[23] No substantial gain in simplicity or accuracy was achieved until Kepler replaced circular orbits by ellipses and postulated a specific kind of *non*-uniform motion (law of areas). Even then, the heliocentric theory seemed to be refuted by the failure of astronomers to detect stellar parallax (apparent motion of stars due to the Earth's motion around the Sun). The initial switch from geocentric to heliocentric astronomy was not based on any discovery of an objective fact but was eventually given a rational justification long after the death of Copernicus. Nevertheless we correctly give him the credit for a major advance in science.

The other example is more controversial but deserves mention because of its great social importance and relevance to present-day adolescents. When Lewis Terman tried out the first version of his intelligence test on white California schoolchildren, he found that the average score for girls was a little higher than for boys of the same age. This result was inconsistent with his preconception that males are at least as smart as females. So he balanced the test items on which boys tended to do better against those on which girls do better, in such a way that the average score would be 100 for each sex.[24] But when Terman and other psychologists administered the test to blacks and found that they scored several points below whites on the average, they did not make a similar revision to equalize the average scores; they were content with the 'discovery' that whites are smarter than blacks. The result is the Stanford–Binet IQ test, widely accepted as an objective measure of inherent intellectual capacity. From it was derived the Scholastic Aptitude Test, used to control admission into selective colleges and hence into desirable positions in American society. Clearly the IQ and the SAT score are socially constructed measures of human minds; the extent to which they may be shown to have objective reality is still unsettled. Certainly their use has had enormous consequences for race relations and for the lives of many people whose academic careers have been affected by college admissions decisions. I suggest that one cannot usefully debate this issue without knowing some of its history.

3   Women and minorities have made important contributions to science

even though outright discrimination and negative social factors have
kept their numbers small. More careful study of the history of science
can ensure that these contributions are not forgotten and that appropriate
role models are available to inspire the next generation. Despite the
considerable efforts that have been made in this area during the past
15 years, science textbooks continue to focus on only one or two
superstars like Madame Curie while overlooking other important women
and minority scientists; or else they segregate them in lists and picture
spreads at the beginning of the book while failing to integrate their
contributions into the appropriate places in the text itself. Often the
contributions themselves are mentioned but not attributed, even though
in the same text contributions by white males are frequently attributed.

To illustrate how one could include the contributions of women in a
historical presentation of physics, here are a few examples of episodes
that would naturally belong in such a presentation whether or not
women were involved.[25]

a   After Newton established the basic laws of mechanics, his comprehen-
    sive treatise, the *Principia* (1687), had a widespread influence
    throughout the world. Although some scientists could use the
    original Latin version, most had to rely on popularizations and
    translations, for example into French (Gabrielle-Emilie du Chatelet).
b   In the early nineteenth century, the latest discoveries in celestial
    mechanics, electricity, magnetism, and heat were explained to the
    scientific public in popular books, thus preparing the way for
    acceptance of the next grand generalization, the law of conservation
    of energy (Mary Somerville).
c   At the beginning of the twentieth century, Becquerel's discovery of
    radioactivity (1896) became the means for a radical transformation
    of atomic physics and chemistry when the intensely radioactive
    element radium was isolated and its properties determined (Marie
    Curie); direct results included the transmutation of elements and
    the nuclear model of the atom.
d   Following the discovery of artificially induced radioactivity (Irène
    Joliot-Curie), research by Hahn and Strassmann led to the
    understanding that a process of nuclear fission of heavy elements
    is possible (Lise Meitner). The development of the atomic bomb
    followed soon afterwards.
e   The structure and properties of the atomic nucleus were successfully
    explained by a 'shell model' (Maria Goeppert Mayer).

f   The suggestion that parity may not be conserved in nuclear reactions involving weak interactions (such as beta decay) was confirmed by experiment (Chien-Shiung Wu), thus launching a new phase in the theory of elementary particles.

g   Fundamental theories in physics were found to depend on a theorem relating conservation laws and symmetry transformations (Emmy Noether).

These proposals for using history in the teaching of science are by no means original. My thesis is that we now have enough concrete evidence showing that the historical approach does have an impact on students, without interfering with their learning of the technical content of science, to proceed further in implementing such proposals.

Acknowledgements

This paper is based on a talk to the Science and Engineering Education Seminar at the National Science Foundation, 8 July 1986. I thank the Directors of Harvard Project Physics (G. Holton, F.J. Rutherford, and F.G. Watson) for providing some of the information used here. My research on the history of science has been supported by grants from NSF (History and Philosophy of Science Program) and the National Endowment for the Humanities (Division of Education Programs).

NOTES

1   In 1986 the most relevant authority to quote on this point is W.J. Bennett, who said in a 1984 report on the findings of a group of prominent educators: '... the study group thought that all students should study the history of science and technology.' (Bennett (1984), p. 9). For recent views on ways to implement such proposals see Bevilacqua and Kennedy (1983), Dhombres (1981), and Home (1983). See also note 16.

2   The term alludes to the phrase 'Whig interpretation of the history of science', used to characterize old-fashioned treatments (usually written by scientists) that judged and presented earlier science in relation to modern science. Thus in describing episodes such as the transition from geocentric to heliocentric astronomy or from phlogiston to oxygen theory in chemistry, the advocates of the view *now* considered correct would be pictured as heroes advancing the progress of science, while the advocates of the view now considered wrong would be shown as villains or at best as misguided and incompetent workers. The phrase is taken from Butterfield (1931), where it is used to describe the tendency of political historians to see British history as progress toward Whig

democracy and to judge individuals and events according to whether they helped or hindered this progress.

Professional historians of science who dismiss most works by scientists as 'Whig history' and therefore not worthy of serious consideration overlook the fact that Butterfield himself later changed his views on the value of the Whig interpretation (see Butterfield (1944)), and that other professional historians have raised doubts about whether a completely non-Whiggish history is desirable or even possible. The issue is relevant to science education because most textbooks that use history at all do so in a Whiggish fashion. For further discussion see S.G. Brush (1974), Hull (1979), Kockelmans (1979), Graham (1981), Oldroyd (1980), and Porter (1981).

3   The Center for the History of Physics, located at the Institute of Physics headquarters in New York City, is the oldest established centre. It maintains and indexes archives of manuscripts and photographs, conducts oral history projects, and publishes a free newsletter. For information write to Dr Spencer Weart, Manager, CHP, AIP, 335 East 45th Street, New York, NY 10017, USA.

The American Physical Society created a Division of History of Physics in 1980, and it now has more than 1700 members. Its symposia at APS meetings are among the most popular sessions there. Speakers at these symposia, and persons elected to its Executive Committee, have included several professional historians of science as well as physicists interested in history. Its *History of Physics Newsletter* is free to members; subscriptions are available to nonmembers. Contact Professor A. Wattenberg, Department of Physics, University of Illinois, Urbana, IL 61801, USA.

A Center for History of Chemistry was established in Philadelphia in 1982, and has embarked on an active programme of conferences, exhibits, and other activities. Contact Dr Jeffrey L. Sturchio, CHC, E.F. Smith Hall/D6, 215 South 34th Street, Philadelphia, PA 19104, USA.

A mathematics history centre had some difficulty getting started when its original host withdrew its support (see Kolata (1983)) but is now being re-established at Brown University.

4   Butts (1982).

5   Rutherford, Holton, and Watson (1970) and Holton (1978).

6   The data come from a variety of sources.

1   Students taking the College Board Achievement Test were asked which text they used; by January 1974 the percentage had risen to 14.6 (see note 13 below). In 1976/77, 12% of school districts used Project Physics (9% had used it in prior years), compared to 11% using PSSC (18% in prior years). The same survey found that 10% of the teachers used Project Physics that year (14% in prior years) but only 4% used PSSC (14% in prior years). Commission on Human Resources of the National research Council, 'The State of School Science: A Review of the teaching of Mathematics, Science and Social studies in American Schools, and Recommendations for Improvements', reprinted in *What are the needs in precollege science, mathematics and social science education? Views from the field* (Washington, DC: National Science Foundation report SE-80-9).

2 A 1978 national survey, answered by 246 physics teachers, found that 17% used Project Physics, 14% used PSSC, 53% used *Modern Physics* (published by Holt). American Association of Physics Teachers, 'High School Needs Assessment Completed', *AAPT Announcer* 8, no. 4 (December 1978), pp. 1065–9.

3 A 1983 survey of 165 schools in one state: 22% Project Physics, 30% PSSC, 30% *Modern Physics* (see Lingenfeld (1985)).

4 An American Institute of Physics survey found that 12% of the undergraduates who earned bachelor degrees in 1983–4 had taken Project Physics and 25% had taken PSSC (see Pallrand and Lingenfeld (1985)).

7 In addition to the survey cited in note 6(1) see Hegelson, Blosser, and Howe (1977), Welch (1979), and Crane (1975).

8 Teachers in New York state reported that, aside from changes in science requirements, the most important factor in changing physics enrollments in their schools was the introduction of a new curriculum, and 'the new curriculum mentioned far more often than any other was Harvard Project Physics'. But the teachers who said this did not claim a significant increase in their own enrollments (see Ehrlich (1977)).

9 Ahlgren and Walberg (1973), Welch (1973).

10 Aikenhead (1974).

11 Quattropani (1978).

12 Lehrman (1982).

13 G. Will Pfeiffenberger (Education Testing Service), letter to Fletcher Watson, November 1974, quoted in *Project Physics Newsletter* (1975), p. 5. Here are the mean scores; 'traditional' means students who said they used a text other than Project Physics or PSSC.

Date	Project Physics students	Traditional students	All students	% taking Project Physics
May 1970	608	608	607	3.6
Jan 1971	570	583	580	5.5
May 1971	599	603	599	7.3
Jan 1972	578	597	589	8.7
May 1972	602	609	605	11.6
Jan 1973	592	597	591	11.0
Jan 1974	585	592	585	14.6

14 F.J. Rutherford, private communication.

15 Members of the History of Science Society are listed in the *Isis Guide to the History of Science*, published every three or four years, with a geographical index and information about graduate programmes. Contact the HSS Publication Office, 215 South 34th Street/D6, Philadelphia, PA 19104, USA.

16 Donovan et al. (1978), Finoocchiaro (1980), Finoocchiaro et al. (1982), and Goldberg et al. (1982). For information about the Committee's current activities

contact Professor Kathryn Olesko, Department of History, Georgetown University, Washington, DC 20057, USA.

17 S.G. Brush (1987).
18 S.G. Brush (1969, 1979, 1980), S.G. Brush and King (1972), Siegel (1979), Kauffman (1979), and Hetherington (1982).
19 Russell (1981).
20 'Neuroscientists are beginning to suspect that everything that makes people human is no more than an interaction of chemicals and electricity inside the labyrinthine folds of the brain', from *Newsweek*, 7 February 1983 (see Jones (1986)). Jones, who teaches physics at the University of Minnesota, is the latest in a long line of commentators pleading for a softening of the harsh image that the scientist has acquired over the last few centuries. One way to do this, as is often suggested, is to teach science historically; see for example L. Brush (1979).
21 Feyerabend (1975), Collins (1985), and Bloor (1976, 1978); 'social construction' interpretations of the history of science are attacked in Roll-Hansen (1983) and Robinson (1986).
22 Holton (1974), Shapere et al. (1986).
23 Gingerich (1975).
24 Terman (1916), Terman and Merrill (1937).
25 For a readable and comprehensive reference work see Ogilvie (1986). Information on contributions of women in the twentieth century may be found in S.G. Brush (1985). Note that the women mentioned in the text were in most cases not the only ones associated with those contributions.

Experiment, Science Teaching, and the New History and Sociology of Science

Harry Collins and Steven Shapin

Introduction

In the late 1940s, reflecting upon the place of science in a liberal democracy and upon public understanding of science in the aftermath of the atomic bomb, the President of Harvard University James B. Conant concluded that a radically new approach to the teaching of science was required.[1] He argued that a proper understanding of what he called 'the tactics and strategy of science' was an essential component in the education of the future citizen. Turning programmatic statements into pedagogic practice, Conant and his colleagues soon produced the celebrated *Harvard Case Histories in Experimental Science*, modelled partly upon teaching materials employed in the Harvard Business School.[2] For Conant a realistic understanding of experimental practices was vitally important for the student who was not going to acquire this understanding directly, at the scientist's bench. For democracy to function in an increasingly scientific and technological society, the citizen had to be put in a position where he or she could not merely see the scientist's product but also grasp the experimental means by which scientific knowledge was generated and evaluated. Without such information the citizen, however 'highly educated and intelligent', 'will almost always fail to grasp the essentials in a discussion that takes place among scientists'; this will be because of 'his fundamental ignorance of what science can or cannot accomplish'.[3] A citizen who is ignorant in this way, Conant argued, was abdicating his rights of citizenship. Thirty-five years later, Conant's analysis is even more relevant. The vast resources demanded for the support of Big Science, the technical sophistication required for comprehending environmental and military issues, the proliferating spectacle of scientific experts disagreeing in the media and the law-courts, the impact of the social sciences in social policy, and, of course, the sheer scale and pace of technological change and its influence on citizens' lives, all call upon educational institutions and teachers to respond anew to Conant's aims. However, there is no reasonable doubt that the education of the citizenry in 'the tactics and strategy of science' has failed to keep pace with the developments that most require that education.

While the situation Conant described has become more pressing, and while his educational initiative has had only limited impact, the conceptual resources available for addressing this problem have become more refined and more suitable for the task. Conant concluded that the best way to give the student a sense of the actual nature of science was through a selective exposure to its *history*: a history which depicted not an idealization of science but scientific practice as it really was and as it had really developed as a human activity and as a part of culture.[4] At the time Conant compiled the *Harvard Case Histories* the academic study of the history of science was in its infancy: hagiographic history dominated, philosophers' idealizations of 'the scientific method' strongly coloured historians' accounts, and the discipline called the sociology of science, to the extent it existed, was concerned with the normative structure of the scientific community rather than with scientific practice.

Within the last decade or so this state of affairs has changed out of all recognition. Interestingly, one of the seminal texts which has helped to change our understanding of the nature of science arose from the pedagogical initiative Conant set in motion. In 1948, the same year in which the *Harvard Case Histories* appeared, Thomas S. Kuhn became a Junior Fellow at Harvard University. While there, Kuhn came into contact with Conant and ultimately helped to teach the course in general science for which the *Case Histories* was the text.[5] Kuhn's *Structure of Scientific Revolutions* (1962) acknowledges the importance of that experience, an importance reflected in the extent to which Kuhn's exemplars derive from the *Case Histories*.[6]

In the two decades since Kuhn's *Structure*, and particularly since the development over the past ten years of a serious sociology of scientific knowledge in Britain and on the Continent, a 'new history and sociology of science' (NHS) literature has come into being. The existence of the NHS literature creates problems and opportunities for those concerned with the teaching of science. Before its availability, there was a rough 'fit' between the idealizations of science found, on the one hand, in the typical science text book and, on the other, in the accounts of the historian and the philosopher of science. This is no longer the case. Now there is a radical misfit between the picture of scientific practice found in the typical text-book and that made available in the NHS literature. This means that those who help to shape the public's understanding of science have not only the opportunity but also the obligation to choose the model of science they present.[7]

The New History and Sociology of Science and Science Education

An understanding of the nature and status of experiment is fundamental to NHS. We cannot here rehearse the details of this understanding, but several of the key points can be briefly set forth.

1 Experiment and its role in the production of scientific truth becomes a problem for inquiry rather than self-evident and given.
2 Experimental findings are seen as inherently defeasible: all experimental findings may be criticized, and no experimental finding need to be taken as a crucial confirmation or disconfirmation of a theory it is said to test.
3 It follows that judgements on the basis of experiment are open-ended in character. Decisions about the status of experimental findings are not dictated by the findings: scientists' judgements may be informed by commitments to certain accounts of what the natural world contains, investments in theoretical and technical resources, and by a range of considerations, some of which are usually thought of as 'external' to science.
4 An 'invention' model of scientific discovery is preferred to 'discovery' model. Through experiment and other means scientists construct their natural world.
5 The transmission of skills such as those involved in experiment follows a 'craft' pattern; the sufficiency of purely verbal means to transmit such skills is cast in doubt.
6 It follows from the above points that the maintenance of consensus over what are taken to be features of the natural world is a social accomplishment.[8]

What is the teacher to do with this NHS literature? The choice would seem to involve educational aims and a decision about whether 'radical' or more 'reformist' innovations in science education are desirable. A number of leading writers in the field of science education evidently contemplate a quite limited role for the history of science. For example, Professor A.P. French aims to use history of science to 'humanize' science education, while retaining its existing goals, structures, and materials.[9] He points to the real constraints posed by the existing curriculum and, especially, to the lack of *time* available for more ambitious changes. The teacher is advised, however, to make sure that 'humanization' is secured by bringing into science education good history rather than bad history. Given these reformist goals, there is nothing special that NHS has to offer: doubtless 'good' history can be found within this literature as much as it can be found within more traditional accounts, but it is hard to see what role could be played by the particular understanding of science offered by NHS. Moreover, it could well be conceded that the text-book account of scientific practice *is* an idealization, but that such idealizations have an important pedagogical function in the training of future scientists.

In contrast to French, we suggest that the potential for the pedagogical use of the NHS literature resides within the framework of more radical goals and, especially, in a radical rethinking of the aims of education *non-scientists* – and educating scientists for their roles outside the laboratory – that is, for educating the citizenry in matters scientific. With Conant, we consider that educating the citizenry in the nature of science is of paramount importance, and we suggest some ways in which NHS can contribute to that end. Our remarks about the aims of science education, because they are largely (but not entirely) directed towards the education of the non-specialist, are mainly concerned with science teaching in school and in the early part of the university curriculum. We do not aim to define the audience more rigidly than that, especially as we recognize that national systems of science education differ enormously.

Let us start by considering the aims of ordinary science education – say those of a school teacher entering a class-room at 10.30 on a dull December morning. These aims may not accurately describe the motives or desires of any particular teacher, but they may be regarded as 'middle-range' ambitions such a teacher might plausibly have, given the constraints presented by typical educational institutions set within our society. Science education can, in this sense, be thought of as having four aims:

1   to enable children to pass examinations in science subjects;
2   to show future citizens the nature of science;
3   to begin to teach future scientists how to do science;
4   to teach children some features of the natural world.

Under the received model of scientific rationality there is a reasonable chance that these four aims can coincide. The crucial characteristic of the received model is the contention that unbiassed methods of investigation will reveal unambiguous, unique, and repeatable true facts about the natural world. Thus, aims 2, 3, and 4 ought to coincide as follows: in learning to do science properly (3), children would uncover true features of the natural world (4), and would learn, in passing, the nature of science, i.e., that careful, unbiassed experiments are repeatable and reveal true features of the natural world (2). To make this work out, it is necessary to design a syllabus that would enable children to find out these things, and 'discovery methods' seem appropriate so long as they can be well enough designed, organized, and managed so that aims 1 and 4 do not become completely lost in classroom turmoil. The tension between discovery methods and more rigid methods of teaching could be seen as the tension between emphasis on aims 2 and 3, the process, and aims 1 and 4, which have more to do with the 'facts'. However, given enough attention to design and organization, 2, 3, and 4 could be made to coincide and aim 1 could be brought into line by providing examinations which reflected the syllabus.

NHS differs from the received model principally in rejecting the view that unbiassed and competent observation and experimentation is enough to reveal unambiguous, repeatable, true facts about the natural world. The new view holds that how something comes to be seen as true, unambiguous and repeatable, is a social process and, therefore, that it is a matter for empirical study. The 'same' region of the natural world may seem to precipitate different facts when observed at different times or places even though similar scientific methods are used. That there is, at any given period, broad consensus within the scientific community about the make-up of the natural world is to be taken as a feature of the social organization of science rather than as a passive reflection of nature. From time to time these broad consensuses are overturned in a wholesale way. From this perspective it is unsurprising that there are many instances of breakdown of consensus or, more properly, it is unsurprising that scientists often disagree about emerging areas of research.

A corollary of this model is that science comprises more than one sort of activity. It comprises work within a broad consensus (normal science), periods when whole traditions of scientific work are being overturned (revolutionary science), and a state in between these two, where controversies arise over new interpretations or findings which cannot be quickly fitted into the prevailing consensus, but which do not promise wholesale revolution: we call this 'extraordinary science'.

If we return to the four aims of science education with the new model in mind, we find that coincidence between the aims is no longer straightforwardly realizable: this is because aims 2, 3, and 4 are all ambiguous. As far as aim 2 is concerned, we need to ask 'What sort of science are we to show our students in order to teach them the fundamental nature of scientific activity?'. Likewise for aim 3 – what are apprentice scientists to be taught to cope with, consensus or confusion? As regards aim 4 we probably want to teach aspects of the current consensus, but do we teach them *as* features of a social consensus or as immutable facts of nature? The answer to this question is not separable from the answers to the other two.

The Choices and Constraints

Normal science

Normal science is science conducted within the consensus. A characteristic of experiments within normal science is that a permissible, fairly narrow, range of outcomes can be specified in advance. Experiments within normal science, like every other experiment or technical accomplishment, require the exercise of technical skills. Skills such as these take time to learn and it is not easy to tell whether or not they have been properly mastered by direct examination. Thus, it is very important that normal science experiments have a restricted range of foreseeable outcomes, because achievement of a result within this range indicates attainment of the appropriate skill. An experimental run which produces anomalous results is accounted, and assessed as, a *failed* run. This retrospective assessment procedure is one of the ways that consensus is maintained in normal science: anomalous results are excluded by fiat.[10] (What consequences this procedure has for 'class-room ethics' is an interesting question which might be discussed elsewhere.)

Extraordinary science

Extraordinary science explores regions in which a well defined range of experimental outcomes is not available. An example of extraordinary science is the detection of gravitational radiation. The question here is not 'how much gravitational radiation is there?' but rather, 'is gravitational radiation detectable on earth?'. Under circumstances such as these it is often unclear whether a scientist has developed the skills appropriate to do the experiment in question successfully.[11] The scientist is not able to refer to the acceptable range of experimental outcomes in order to determine if his experiment should be accounted a success or a failure. Thus, scientists may disagree about who has developed a successful experiment, e.g., is it those who have detected gravitational radiation or is it those who have not? Science of this sort is confused, unclear, indecisive, and often characterized by bitter personal dispute, and the controversies it engenders are often long-running.[12] Under these conditions a failure of expertise to bring about consensus over the constituents of the natural world is very evident and the received model of scientific rationality is most clearly wanting.

The student as potential scientist

Nearly all science is 'normal science'. In so far as the concern of the science educator is to begin the training of scientists, then it is the abilities associated with the practice of normal science that ought to be taught.[13] Thus, when the student does an experiment which produces an anomalous result (a result outside the range of normal expectations), the student *should* be told that he or she has failed to perform the experiment properly and asked to continue until a suitable result is achieved. The production of a suitable result should be celebrated as a vindication of the consensual view of nature and of the received view of scientific knowledge. *It is belief in the received model* that makes scientists persevere in the face of recalcitrant results. The interpretation of an anomalous result as a consequence of the experimenter's errors, and the strong desire to produce the correct result, is the practical working out of the received model of scientific rationality: it is the practical expression of what may (non-pejoratively) be called 'the ideology of science'.

Thus, one might plausibly argue that there is little reason to effect changes (other than cosmetic ones) in the education of future specialists. Nevertheless, several features of the understanding of science offered by NHS could be considered even here. For instance, sociologists of science have demonstrated that experimental skills are not in fact reliably transmitted by purely verbal means.[14] In actual scientific research it proves necessary to acquire such skills by 'tacit' means: by apprenticeship, by watching competent experimentalists perform, or by the processes colloquially called 'sitting by Nellie'. The techniques by which the making of puff-pastry is taught to novice cooks are the same by which the performance of a titration is taught to novice chemists. Since this is the case, science teachers might reflect whether excessive reliance is not being placed on the capacity of written instructions to transmit desired experimental competences and see whether changes in the organization of laboratory teaching might result in more effective transmission of such skills. At a higher level, the future administrators of the science budget ought to be able to mount an informed defence of the need for travel to conferences, for exchange fellowships, and for other forms of international co-operation.

Then, it has been suggested that there often exists a sort of 'trauma' attending the passage from the state of being a science student to the state of being a scientific researcher.[15] If the science student is wholly trained in a situation in which the answers are pre-given in the questions, he or she may experience disorientation when confronted by the real life of many scientists – a life in which the answers may not be known in advance, in which (particularly in extraordinary science) the existence of natural phenomena cannot be decided apart from judgments about whether or not an experiment has been competently performed. From time to time there has been concern over 'drop-out' or 'wastage' at this stage of the production of scientists – a wastage which might conceivably be diminished by exposing science students more effectively to the contingencies affecting actual scientific experimentation, including the social processes by which scientific consensus is made. These points might be argued at length, but there is less ambiguity when it comes to thinking of science education for the ordinary citizen.

The student as potential citizen

The citizen, and the democratic society of which he or she is to become a competent member, is ill served if education provides only the ideology of science. The most important reason for saying this involves the problem of 'anti-science' attitudes. As we have indicated, disagreements among scientific experts is an increasingly common feature of modern scientific and technological society. Experts disagree and citizens (and the institutions which are said to be answerable to them) are asked to react and to decide. How will the citizen react to the spectacle of disagreeing experts if he or she is offered only the ideology of science? Arguably, the reaction will be polarized. In the ideological version, science is either wholly true or wholly baseless. If this is what is believed, then disagreement among experts will generate profound disillusion about what is taken to be science *tout court*. Scientists, it will be thought, are all incompetents, or liars, or intellectuals available for hire to powerful interest groups. There will be no shades of opinion, no feeling that disagreement is natural to much of science, no tendency, that is, to cope with the spectacle of disagreeing experts as an endemic feature of 'good' science. Thus, the ideological version of science, so often believed to be the best way of guaranteeing proper respect for science in a democratic society, may actually be responsible for much of the anti-science attitude that concerns scientists and science teachers. The NHS perspective, contrarily, may assist the future citizen in regarding scientists as 'the best possible experts', while not generating the unreasonable expectations and consequent polarizations which are responsible for anti-science opinions.

Science teaching has the potential to provide the future citizen with an alternative account of the nature of science to that offered by scientists themselves.[16] This is the crux of our suggestion for reform. Anomalous results, which must be immediately reinterpreted as experimental errors if the ideology is to be properly internalized, could, on the other hand, be taken as typical outcomes of experiment in frontier areas (extraordinary science). It is the very *disorganization* of the discovery method which would have the greatest heuristic value in this case. The student could be shown the possibilities for disorganization and disagreement where experts are asked to comment upon matters at the limit of their expertise in areas where consensus is not yet formed. Demonstrating this would become the point of the discovery method.

Moreover, the next step, the process of the formation of consensus from this potential chaos, could be practically revealed in the class-room. After all, the students, when they are being taught as potential scientists, are being shown how to reduce disorder to order in their practice as informed by the received model of scientific rationality.

It is clear, however, that there is a tension between teaching the student as potential scientist and teaching the student as potential citizen. Subtle compartmentalization is needed to grasp both viewpoints.

The teacher as scientist

So far as the current state of education is concerned, teachers of science (at least in Britain and the United States) will have had no opportunity to experience science in other than its ideological guise. Only a few scientists – those who have become involved in scientific controversy – will have discovered the non-ideological face of science. In fact, most science teachers will have had little experience of science outside their own education, and so will have had nothing but an unrelieved diet of the ideology. It is, of course, the successful absorption and application of the received model which leads to academic success and, indeed, to scientific success at all but the very top levels. We can take it, therefore, that science teachers are successful absorbers and appliers of the received model of scientific rationality, and we may suppose that the same applies to those who design syllabuses. In this respect there is no difference between those who favour the old rote learning methods and those who favour discovery methods. Discovery methods are merely a more subtle, and probably a more effective, tool for teaching practical applications of the received model.

The Four Aims and a Suggestion for Educational Experiment

Let us now turn back to the four aims of science education mentioned above. These were:

1   to enable children to pass examinations in science subjects;
2   to show future citizens the nature of science;
3   to begin to teach future scientists how to do science;
4   to teach children some features of the natural world.

We can see that 1, 3, and 4 can be made to coincide under the new model provided that the purpose is to teach science in an ideological way. However, this is entirely inappropriate as regards aim 2. Aims 2 and 4 can be brought into alignment if features of the natural world are treated as features of the current socially organized consensus, and the experimental method can be used to good effect so long as it is the social organization of consensus that is the focus of attention rather than an abstract idea of the proper outcome of experiments. This requires that the achievement of consensus in the classroom be taken as a microcosm of the achievement of consensus – over a much larger time period – in the scientific community. Now and again, we suggest, students' attention might be drawn to the way the class, under the direction of the teacher, reduces an initially disordered set of quasi-findings into experimental support for a favoured (the correct!) hypothesis. The parallels between the classroom lesson and the life histories of scientific controversies, as they move toward closure, are striking.[17]

So far as aim 1, passing examinations, is concerned, under the new model it could only coincide with the other aims if some part of the syllabus were devoted to questions about the social organization of science – a topic currently thought of as 'Science Studies'.

Recommendations and Problems

Although these are clearly very early days for the project of incorporating NHS into innovatory science teaching, some provisional practical recommendations may be offered. First, the time is ripe for a new teaching resource along the lines of the *Harvard Case Histories* developed thirty-five years ago. Just as Conant drew upon the best historical accounts of the actual nature of science available to him, so we can now do the same, drawing upon the radically different picture of science provided in the NHS literature. However, our aims in doing this will be very much the same as Conant's and our target will be very much the same as that envisioned by him: the group of students somewhere between the last years of school and the early years of university education.[18] Second, in a less concrete way, we might suggest that the actual teaching of experiments could (at least occasionally) be turned into an object of study and reflection. Along with teaching students the appropriate skills and aspects of the natural world, we could make the

teaching of experiment into an object lesson in the processes by which
scientific skills are transmitted and social consensus about reality is
produced. This, however, would be an ambitious exercise and we do
not know exactly how it should be carried out.

Finally, a number of problems concerning the approach recommended
in this paper have been pointed out to us. One needs very carefully to
consider the educational setting in which such innovations are to be
introduced. As we have said, much of what we recommend applies
to late-school and early-university teaching. Depending upon the
institutional environment, school may provide very different conditions
from university, and it might be argued that such changes are wholly
inappropriate for school or whollly inappropriate for university.
Moreover, we have not considered the enormous differences that obtain
in various national systems of education. Necessarily, our ideas have
been informed by our familiarity with patterns in Great Britain and in
the United States. Structures and aims which characterize other national
systems may offer further opportunities as well as further constraints to
innovation along these lines.

This is an amended version of a paper originally given to a conference
on 'Using the History of Science in Innovatory Physics Education',
Pavia, Italy, 5–9 September 1983. A facsimile of the typescript of the
original was reproduced in the proceedings of the conference. It was
also translated into German and published in *Physica Didactica*, **33**(11),
(1984), 33–46.

NOTES

1 Conant (1947).
2 Conant (1948); the model of business case studies is acknowledged in Conant
   (1947).
3 Conant (1948), vol. 1, p. vii; see also Conant (1967), pp. 8–12.
4 Conant (1967), especially pp. 6–9.
5 For an account of Kuhn's relationship with Conant see Merton (1977),
   pp. 3–141and 81–9.
6 See the preface to Kuhn (1962).
7 Reviews of selected NHS studies are Barnes and Edge (1982), Shapin (1982),
   and Collins (1982).
8 These points are developed in depth and case studies are discussed in Collins
   (1985).
9 French (1983).

10 T.S. Kuhn's 'The Function of Dogma in Scientific Research', in Crombie (1963), pp. 347–69.
11 Collins (1985).
12 See also Farley and Geison (1974) and Shapin (1979), pp. 139–78.
13 For an alternative treatment of school students as scientists see Ros Driver's work, for example Driver, Guesne, and Tiberghien (1985) and Driver (1983).
14 See, for example, Collins (1974) and Collins and Harrison (1975).
15 For example, by Professor A.P. French of MIT (French 1983). For a broad coincidence with the outlook presented here, see Edge and Aikenhead (1983).
16 For a detailed discussion of the way in which science is portrayed on television as a body of 'certain knowledge' rather than as a 'knowledge-producing activity', see Collins (1987).
17 For a fascinating account of the development and maintenance of cognitive order in science lessons using the 'discovery' style of teaching, see Atkinson and Delamont (1977).
18 A book with this aim in view, *Science in Practice* by H.M. Collins and R.H. Millar, is currently in preparation for Cambridge University Press.

The Historiography of Science Retrospect and Future Challenge

Richard Jones

The Traditional Approach

Until recently, most histories of science gave the impression that science is not really a historical phenomenon at all. They were based on the assumption that the history of science consisted of the gradual revelation of *the* truth about the natural world. This being the case, historical events could speed up or slow down science's progress, but they could play no part in determining either the general direction of this progress or the nature of the truth which was eventually revealed. In Karl Figlio's words, science was presented as if it were 'more a discovery than a product'.[1]

This traditional approach to the history of science was partly due to the influence of practising scientists who dominated the discipline during its formative years. Most 'practitioners' history aimed to validate contemporary scientific theory and practice by demonstrating the existence of an unambiguous scientific path, leading from the ignorance of the past to the enlightenment of the present. As Arnold Thackray says of American history of science during the first half of the twentieth century, it 'displayed the heroic achievements of great scientists of the past' and constituted 'a possible basis for confidence in the continuation of the achievements of scientists in the future'.[2] But this approach was also due to the widespread acceptance of the nineteenth century positivist view of science. As far as positivists were concerned, science provided the only source of genuine knowledge – 'the scientific method' constituting a kind of 'royal road' to truth. They also believed that human progress was guaranteed by the inevitability of scientific progress. This emphasis upon the steady accumulation of scientific knowledge due to the application of an objective scientific method was, until recently, the most widely accepted view of science. It therefore provided a 'sound' philosophical foundation for histories of science which demonstrated the gradual discovery of the truth about the natural world.

The adoption of this 'ideology of science' by traditional historians has ensured that most histories of science are written from the perspective of present-day knowledge. In other words, the scientific knowledge of the past is evaluated by comparing it with the scientific knowledge of the present, and the amount of attention which a particular concept or theory receives depends upon the extent to which it can be placed in a casual sequence which culminates in modern science. In François Jacob's words, the historian is involved in

... a search for the thread which guided thought along the path of current theories. This is reverse history so to speak, which moves back from the present towards the past. Step by step, the forerunner of the current hypothesis is chosen, then the forerunner of the forerunner, and so on.[3]

This approach distorts the history of science because it causes undue emphasis to be placed upon those theories which can be 'linked' to modern science, with other theories, which may have been of equal or greater importance at the time, being rejected as error. Thus Erik Nordenskiold, having devoted all of one sentence of his classic *The History of Biology* (1928) to Paracelsus' advocacy of the doctrine of 'signatures', curtailed his account with the following remark: 'To go deeper into these fantastic ideas is hardly worthwhile; what has already been stated may seem to many readers more than enough of such nonsense.' He then went on to point out that, despite this, Paracelsus 'was in many respects a pioneer' because he 'did his best to treat wounds hygienically' and many of his methods – such as the treatment of syphilis with mercury – 'still hold good'. By applying scientific criteria which were acceptable in the 1920s, Nordenskiold managed to 'link' Paracelsus' work to his (i.e. Nordenskiold's) present, but only at the cost of rejecting most of the views which Paracelsus actually held. In addition, Nordenskiold totally ignored the widespread influence which 'such nonsense' had throughout the sixteenth and seventeenth centuries.[4]

The distortions which are created when this approach is adopted can be far more subtle. Take, for example, traditional accounts of eighteenth century botany. As Francois Delaporte points out, these tend to rest on two preconceived notions:

The first is that the detection of analogies was a cause of stagnation; the second that only a method based on observation and experimentation could bring progress in vegetable physiology . . . The analogical method was a hindrance, it is argued, because it was a procedure the eighteenth century presumably inherited from the Renaissance or even Antiquity . . . In contrast, the work done by experimentalists and observers was supposedly a prefiguration of the nineteenth century.[5]

This kind of analysis is clearly apparent in P.C. Ritterbush's *Overtures to Biology: The Speculations of Eighteenth Century Naturalists* (1964). Here, having referred to analogies between plants and minerals as 'a curse several times over', Ritterbush goes on to state that the analogy of circulation between plants and animals eventually 'fell into disfavour, except among speculators who cared but little for the truth'. In doing so, he divided botanical knowledge into two separate domains – the analogical and the experimental – when no such separation actually existed at the time. Most experimentalists made use of analogy and many analogists were involved in observation and experiment.

These distortions become even more serious when based upon the historian's perceptions of the fundamental conceptual structure of contemporary science. Thus Ernst Mayr dismisses much of *The History of Biology* because it was written at a time when 'Darwinism had reached its nadir of prestige, at least on the European continent'. This 'anti-Darwinian bias' ensured that Nordenskiold's 'presentation of the history of evolutionary biology is virtually useless.'[6] Yet Mayr leaves himself open to a similar critique when he states: The main reason . . . why histories are in constant need of revision is that at any given time they merely reflect the present state of understanding; they depend on how the author interpreted the current *Zeitgeist* of biology and on his own conceptual framework and background. Thus, by necessity the writing of history is subjective and ephemeral. Although this statement does not make it clear whether such subjectivity is to be resisted or embraced, Mayr goes on to leave the reader in no doubt about his historiographical outlook: I feel that the study of the history of a field is the best way of acquiring an understanding of its concepts. Only by going over the hard way by which these concepts were worked out – by learning all the earlier wrong assumptions that had to be refuted one by one, in other words, by learning all past mistakes – can one hope to acquire a really thorough and sound understanding. It is this approach to history which encourages Mayr to describe Oken, Schelling, and Carus as authors whose 'silly constructions the modern reader can only read with embarrassment.'[7]

Mayr's belief that the history of 'biology' should serve the needs of contemporary biology highlights the most fundamental form of distortion associated with the traditional approach to the history of science. This concerns the imposition of modern disciplinary boundaries upon past forms of knowledge. As Roy Porter comments:

It has been all too common to take for granted the nature and distribution of present scientific disciplines, their contents and their parameters, as if they timelessly reflected nature. Yet, the reverse is true. That disciplines are, on the contrary, historical and cultural products is evident from the fact that terms such as 'biology' and 'geology', that are currently taken for granted, were coined scarcely two centuries ago.[8]

Yet it is this very recognition that present scientific disciplines are relatively recent cultural products which is masked when historians of science impose modern disciplinary boundaries upon past forms of knowledge. This problem even extends to the use of the term 'science' – particularly when historians apply it to ancient times. As Rupert Hall points out, 'using a word of modern currency for which the Greeks possessed no equivalent, such a title as 'Greek Science in Antiquity' is scarcely less anachronistic than a study of 'Cubism in Antiquity' would be'.[9]

The Loss of Certainty

As early as the 1920s there were already strong indications that the traditional approach to the history of science was no longer viable. Of particular significance were the developments in theoretical physics during the early decades of the twentieth century. Einstein's work on relativity totally undermined the conceptual foundations upon which an understanding of the physical world had been based for well over two hundred years. In doing so, it convinced a number of philosophers that positivist ideas concerning the gradual accumulation of scientific knowledge no longer provided the basis for a convincing model of scientific change. The emergence of quantum theory was equally significant. It cast doubt on the notion of causality and seemed to suggest that certain knowledge of the physical world was impossible –

statements of probability being the best that could be hoped for.
Reflecting on the epistemological implications of this revolution in
theoretical physics, Werner Heisenberg wrote:

The old division of the world into objective processes in space and time and the
mind in which these processes are mirrored . . . is no longer a suitable starting
point for our understanding of modern science . . . Science no longer confronts
nature as an objective observer, but sees itself as an actor in [the] interplay between
man and nature. The scientific method of analysing, explaining and classifying has
become conscious of its limitations, which arise out of the fact that by its intervention
science alters and refashions the objects of investigation. In other words, method
and object can no longer be separated.[10]

It is true that positivism – particularly the logical positivism of the
Vienna Circle – continued to dominate the philosophy of science during
this inter-war period, and that many positivists took relativity and
quantum theory as their starting point. But even within the Vienna
Circle there was considerable uncertainty concerning the nature of
scientific truth, with some members favouring the view that it consisted
of a correspondence between logical propositions, not a correspondence
between logical propositions and the real world.

During this same inter-war period major changes were also taking
place in historiography. By the end of the nineteenth century, Rankean
'scientific history' had come to dominate the universities of Europe. Its
practitioners concentrated on the main political and military events of
European history in an attempt to demonstrate the rational evolution
of the nation state. Following the First World War, however, criticisms
of this approach to history turned into a general revolt. As George
Iggers points out, the events of the war years undermined the notion
of progress on which the traditional view of history depended: 'The
idea of progress gave way among thinkers as different as Spengler and
Horkheimer to pessimistic visions of the self-destructive forces inherent
in modern technological societies'.[11] Marc Bloch and Lucien Febvre,
having been impressed by the rejection of certainty in physics, argued
that historians should follow this lead and reject the outmoded certainty
which characterized contemporary Rankean 'scientific history'. They
were particularly critical of the emphasis placed upon short-time frame
descriptions of well documented events. Such 'event history', they
maintained, is almost certain to distort the past, for during the collection
and organizing of a potentially infinite number of facts, the historian
has to be highly selective with regard to the data eventually used. This

ensures that much event history degenerates into the construction of an ideology, or myth, which says more about the historian and his times than it does about the period under study.[12] Bloch and Febvre therefore placed great emphasis upon putting short-term events into the context of relatively stabler long-term structures. They aimed to establish the totality of conditions in which a problem or event is located and placed little emphasis upon the detailed causal schemes of traditional history. This approach is evident in Bloch's claim that:

European feudalism in its most characteristic institutions was not an archaic web of survivals. At a certain moment in our past it was born from a particular total social situation . . . A historical phenomenon can never be fully explained if seen divorced from its background . . . The Arab proverb that men are more similar to their times than their fathers, said this before us.[13]

Viewed from this structural–functional perspective, historical events become functions, and change is seen not in terms of progress or continuity, but in terms of a process of restructuring which creates a need for other functions.[14]

A number of British and American historians also mounted powerful attacks on traditional historiography during this period. One of the most widely quoted attacks was made by Herbert Butterfield in *The Whig Interpretation of History* (1931).

It is part and parcel of the Whig interpretation of history that it studies the past with reference to the present . . . The Whig historian stands on the summit of the twentieth century and organizes his scheme of history from the point of view of his own day. [He searches] for likenesses between past and present, instead of being vigilant for unlikenesses; so that he will find it easy to say that he has seen the present in the past, he will imagine he has seen a 'root' or an 'anticipation' of the twentieth century, when in reality he is in a world of different connotations altogether. . . .[15]

Butterfield's critique was, if anything, even more relevant to histories of science than it was to the general histories at which it was aimed. It was further developed by R.G. Collingwood into an extreme form of historical relativism. In *The Idea of History* (1946), Collingwood argued that during different historical periods, different intellectual presuppositions, interests, concepts, etc. prevail. This being the case, intellectual standards belonging to the twentieth century have no relevance or authority when applied to another period. Such a period can only be understood in terms of its own intellectual standards.

This growing critique of traditional historiography has had a major impact upon the way in which general histories are written. It has encouraged a move away from 'scissors and paste' linear studies, where the past tends to be dealt with as a primitive forerunner of the present, towards broader 'cross-sectional' studies, where the past is dealt with 'on its own terms'. This in turn has encouraged a move away from an almost exclusive concern with political and military events towards a greater concern with the wider social context. As Lloyd S. Kramer has recently pointed out: 'Although the debate about historical methodologies never leads to a final consensus, it seems that social history now provides the most influential approach within contemporary historiography; that is, a great many historians believe that social history offers the most thorough access to reality.'[16] These developments in science, philosophy and historiography have major implications for the traditional approach to the history of science.

The Discontinuist Approach

The developments outlined above eventually began to have some influence upon the way in which histories of science were written. This was largely due to the increasing involvement of philosophers and historians in the writing of such histories. Many philosophers of science came to believe that certain fundamental questions concerning the epistemology of scientific knowledge could only be adequately approached via actual studies of the way in which science changes through time. As far back as the 1830s, William Whewell had attempted to base his philosophy of science on his reading of the history of science, and both Ernst Mach and Pierre Duhem made philosophical concerns a prime motive for investigating the history of science.

During the 1930s Gaston Bachelard and Alexandre Koyré were the leading advocates of a historical approach to the philosophy of science.[17] As a result of their work, it became increasingly obvious that traditional philosophies of science bore little resemblance to what actually happened in science. They were particularly critical of the positivist belief that science grows due to the steady accumulation of objective scientific discoveries. Like Bloch and Febvre, Bachelard's work was initially inspired by the revolutionary changes which were taking place in theoretical physics. An analysis of these changes convinced him that

There is no transition between Newton's system and Einstein's system. One cannot get from the former to the latter by collecting knowledges, taking double pains with measurement, slightly modifying principles. On the contrary, an effort of total novelty is required.[18]

This led him to the conclusion that the history of scientific thought is essentially discontinuous, that is, it is punctuated by periods of revolutionary change during which the whole conceptual structure of science experiences a fundamental 'mutation'. The fact that Bachelard was writing about discontinuities in the history of science almost thirty years before Thomas Kuhn 'popularized' the notion within Anglo-American scholarship helps to explain Ian Hacking's comment that 'Kuhn was a sensation for us but rather old hat in France'.[19] Bachelard also argued that a radical discontinuity exists between science and everyday experience. In particular, a fundamental difference exists between the objects which emerge from a scientific study of the world – objects such as phlogiston, aether, electrons, gravity waves, etc. – and those which emerge from everyday experience. Bachelard referred to the latter as 'natural' objects because they are directly given to sense experience. Scientific objects on the other hand are 'secondary' objects because they only emerge within a particular theoretical framework (or 'problematic') – a framework which is ultimately destined to be replaced. As Roy Bhaskar points out, the acceptance of Bachelard's epistemology has major consequences for the history and philosophy of science: 'In short, it becomes necessary to distinguish clearly between the unchanging real objects that exist outside the scientific process and the changing cognitive objects that are produced within science as a function of scientific practice.'[20] It also becomes necessary to reject traditional positivist notions concerning the gradual progress of science towards perfect correspondence with the world – whether real or ideal.[21]

Bachelard's references to mutations and epistemological breaks had a considerable influence upon Koyré's attempt – in *Galileo Studies* (1939) – to demonstrate that a radical break occurred between medieval and Galilean physics. They have also had a lasting influence upon the way in which histories of science are written in France. This influence is particularly apparent in the work of Georges Canguilhem who succeeded Bachelard as head of the Institut d'Histoire des Sciences at the University of Paris.[22] A central feature of Canguilhem's work was his rejection of the traditional belief that the objects of our knowledge of the natural world are essentially unchanging. Like Bachelard, he believed that the concepts which come to dominate a particular period play a major part in determining the objects which the scientists of that period seek to study. When a major change occurs in this conceptual structure, the objects of knowledge also change and science takes on a radically different form. Seen from this perspective, the history of science consists of a series of relatively stable conceptual 'spaces' within which human thought is constrained to operate for a period of time. Canguilhem's adoption of this discontinuist outlook led him to attack one of the main objects of traditional histories of science – 'the precursor'. He was particularly impressed by Koyré's comment that: 'It is . . . evident – or at least it ought to be – that no one ever thought of himself as someone else's precursor, and indeed could not have. Hence to consider him as such is the best way to fail to understand him.'[23] But Canguilhem went further than this and claimed that

Strictly speaking, if precursors existed the History of the Sciences would be meaningless, since science itself would only apparently have a historical dimension . . . The willingness to search for, to find and to celebrate precursors is the clearest symptom of an inaptitude for epistemological criticism. Before putting two distances on a road end to end, it is advisable to be sure that it really is the same road.[24]

The continued search for antecedents within the history of science convinced him that 'this science of the past' is 'the past of today's science', a past which 'serves the needs of the day.'[25]

Another French historian who was influenced by Bachelard was Michel Foucault, whose early work constitutes a radical critique of traditional histories of science.[26] Foucault not only rejected the notion of immutable objects of knowledge, he also rejected the notion of an immutable subject of knowledge, that is, he believed that the most intimate aspects of human thought and feeling are subject to radical change during the course of history. This explains why he was even more critical than either Bachelard or Canguilhem of any attempt to establish simple continuities between past and present. Foucault's archaeological approach to history, far from establishing a continuity in human thought, reveals 'discontinuities, ruptures, gaps'. In doing so it unties 'all those knots that historians have patiently tied'.[27] This approach is exemplified by his work on the emergence of biology during the early nineteenth century. Here he states that: 'Historians want to write histories of biology in the eighteenth century; but they do not realize that biology did not exist then, and that the pattern of knowledge that has been familiar to us for a hundred and fifty years, is not valid for a previous period.'[28] What did exist in the eighteenth century was natural history, a form of knowledge which is bound to be misunderstood when presented as a primitive forerunner of modern biology. Although several studies in the history of science have been influenced by this archeological approach, Foucault's later genealogical work – which concentrates on the relationship between knowledge and its socio-political context – has yet to be properly evaluated by historians of science.[29]

In England and America it was developments in historiography, rather than developments in the philosophy of science, which initially had the greatest impact upon the way in which histories of science were written. During the 1950s a number of leading historians of science – for example, Marshall Clagett, A.C. Crombie, Rupert Hall – made a conscious effort to avoid the worst excesses of Whig historiography. Rather than treating the scientific knowledge of the past as if it were a simple anticipation of the scientific knowledge of the present, they made an attempt to understand the past on its own terms. This approach is clearly apparent in A.C. Crombie's claim that: "It is an interpretation of the aims, conceptions and solutions of the past, 'as they occurred in the past', that is the primary quarry of the historian of science.'[30] Despite this claim, however, many historians of science continued to write thoroughly Whiggish histories. This was largely due to their continued acceptance of the traditional positivist view of scientific progress. Even Crombie saw no contradiction in prefacing the statement quoted above with the following pronouncement: 'Unlike other disciplines dealing with the world, the solutions to problems in science, past and present, can be judged by criteria that are in most cases objective, universally accepted and stable from one period to the next.' Attempting to write non-Whiggish histories of science while retaining a traditional positivist outlook is bound to be self-defeating. This helps to explain John Krige's observation that many eminent historians of science 'oscillate between the continuist and discontinuist poles without apparently realizing that they are doing it.'[31]

Within Anglo-American scholarship, a serious challenge to the positivist view of scientific progress did not appear until the publication of Thomas Kuhn's *The Structure of Scientific Revolutions* (1962). Ever since this work appeared, the philosophy of science has become increasingly dominated by historical studies.[32] As Mary Hesse comments, the 'recent revolution in empiricist philosophy of science may be characterized in many ways, not least important of which is the turn from logical models to historical models'.[33] Moreover, the work of writers such as Thomas Kuhn, Imre Lakatos, and Paul Feyerabend has focused on the very problem which concerned Bachelard, Koyré, and Popper in the 1930s, that is, the problem posed by the existence of discontinuities in the history of science.[34] The appearance of Kuhn's model of scientific change ensured that the concepts and theories of science could no longer be regarded as belonging to a single structure which gradually developed due to the accumulation of truth and the elimination of error. On the contrary, the history of science was presented as a series of relatively isolated 'paradigms', each with its own conceptual structure, its own criteria for selecting problems, and its own methods for evaluating theories. Unfortunately, however, Kuhn's original use of the term 'paradigm' was somewhat vague. It shifted between a broad sense (equivalent to a total conceptual structure) and a narrow sense (equivalent to a single theory). In a 'Postscript' to the second edition of *The Structure of Scientific Revolutions* (1969) he conceded this problem and modified his views in such a way as to reduce the incommensurability between two successive paradigms. This reduced the magnitude of the break between one scientific period and another and negated much of the radical epistemological impact of his original thesis.

It is Paul Feyerabend who has taken the relativism inherent in Kuhn's early work to its logical conclusion. Feyerabend totally rejects the traditional attempt to equate scientific change with rational progress. He argues that an infinite number of theories will fit the data provided by our experience of the world (this idea has been known since Duhem as 'the underdetermination of theory by empirical data'). This being the case, the empirical success of an existing theory is no guarantee of its truth. At first glance this may seem to be a rather extreme position. After all, most people assume that the truth of a particular scientific theory has been firmly established when it can be used to make successful predictions – particularly when this leads to technological advance. But the history of science demonstrates the error of this equation. Take, for example, Newtonian mechanics. This continues to be of great pragmatic value and is used by engineers and architects to build machines and structures which work. But from the perspective of modern physics the truth value of Newtonian physics is zero – its concepts of space, time, mass, and force are all wrong, and so are all their deductive consequences. The obvious conclusion is that pragmatic success permits a variety of conceptual structures. The real world may well place limits on the nature of these structures but it does not appear to determine which structure is adopted at any particular time – hence the central question of contemporary history and philosophy of science, that is, what *does* determine the change from one conceptual structure to another? The other main feature of Feyerabend's thought concerns his belief that rival theories are often so incommensurable that it is not possible to formulate the basic conepts of one theory in terms of another. As a result they cannot share any observation statements and they cannot be rationally compared:

The new conceptual system that arises (within relativity theory) does not just deny the existence of classical states of affairs, it does not even permit us to formulate statements expressing such states of affairs. It does not and cannot, share a single statement with its predecessor.[35]

This leads him to the controversial conclusion that competing theories are equally reasonable alternatives, with one being eliminated in favour of another only as a result of subjective choice. 'What remains are aesthetic judgements, judgements of taste, metaphysical prejudices, religious desires, in short, what remains are our subjective wishes'. Given this analysis, Feyerabend is highly critical of the way in which certain theories are turned into petrified dogma, while alternative theories are rejected as error. Hence his claim that the institutionalization of science in schools, universities, and research centres is equivalent to the imposition of an ideology.

The Contextualist Approach

Ever since Karl Mannheim's *Ideology and Utopia* (1936), sociologists have shown an increasing interest in the sociology of knowledge. But it was Robert K. Merton's *Science, Technology and Society in Seventeenth Century England* (1938) which inspired the application of this approach to the history of science. Despite his pioneering study, however, Merton consistently failed to challenge the traditional positivist view that the history of science is the history of the rational accumulation of objective knowledge. The impetus for such a challenge was, of course, provided by Kuhn's early work. This generated a considerable amount of interest during the 1960s because it accounted for shifts from one paradigm to another by reference to the sociology of the scientific community. But, like Merton's early work, it ultimately failed to deliver its radical epistemological or sociological promise. According to R.M. Young, such a failure was inevitable, for there was an inherent problem in Kuhn's analysis:

In the course of finding his insights liberating because of their introduction of social factors in the process of conceptual change, it was not noticed that he thereby excluded socio-economic factors from the substance of science . . . He has provided us with a sort of internalist's contextualism concerned with a social milieu, but it is the social context of scientists in the society of science, not in the world.[36]

It was the narrow 'internalist's' approach to the history of science which dominated contextualist studies until well into the 1970s.

The tendency to limit the scope of the scientist's wider social context has received its most overt methodological expression in Larry Laudan's 'arationality assumption'. This states that the 'sociology of knowledge may step in to explain beliefs if and only if those beliefs cannot be explained in terms of their rational merits.' Laudan goes on to state that historians of science should concentrate upon 'the rational historiography of ideas', that is, the rational reconstruction of objective scientific knowledge. This will leave sociologists to deal with the more peripheral aspects which cannot be explained in rational terms. In this way, a clear 'division of labour' is created 'between the historian of ideas and the sociologist of knowledge'.[37] Although a number of strong cases have been made for the rejection of the arationality assumption,[38] it is more interesting to reflect upon why Laudan felt it necessary to state so explicitly an assumption which was widely accepted – even among sociologists of science. In retrospect one can suggest that the traditional consensus among historians and sociologists of science was rapidly breaking down. Throughout the 1970s there were increasing signs that the traditional constraints within the sociology of science were no longer viable and that the rational reconstruction of the history of science was about to be challenged by a sociological reconstruction which would no longer avoid the analysis of 'objective' scientific knowledge.

The first evidence of a more radical contextualist approach appeared in the late 1960s with the work of a number of historians of science who were influenced by the New Left movement. This was essentially a students' movement inspired by certain neo-Marxist works such as Herbert Marcuse's *One Dimensional Man* (1964). Classical Marxists have always maintained a somewhat ambivalent attitude towards science and the influence of Marxism on Anglo-American histories of science has been slight – notable exceptions being Joseph Needham's *Science and*

Civilisation in China (1954) and J.D. Bernal's *Science in History* (1954).[39]
Within New Left theory, however, scientific and technological rationality
are given a central role in both the emergence and the maintenance of
advanced capitalist societies. One of the first historians of science to be
influenced by this new political perspective was R.M. Young. His early
studies on the similarities between evolutionary theory and social theory
in nineteenth century Britain convinced him that both biologists and
social theorists were influenced by common socio-political concerns.
This 'common context' hypothesis led him to advocate the following
approach:

Rather than having internalists *versus* externalists in the history of science, with
both of these groups relatively separated from other sorts of historians, an approach
might be developed which routinely considers social and political factors in scientific
research and scientific factors in social, economic and political history . . . In a
sense I want to marry history of socio-economic theory and history of biology, or
– to alter the metaphor – to show that Malthus, Paley, Chambers, Darwin, Charles
Lyell, Herbert Spencer, A.R. Wallace, and others were part of a single debate.[40]

Young's belief that scientific knowledge is inextricably intertwined with
the scientist's socio-political context has influenced a number of
contextualist studies – including some by authors who do not share his
political sympathies. More recently, a number of neo-Marxist historians
have attempted to demonstrate that scientific knowledge is capable of
providing covert support for wider socio-political developments.[41] This
approach is apparent in Roger Cooter's study of the popularization of
physiology in early nineteenth century Britain. Here Cooter claims that
the dissemination of physiological knowledge played an important role
in 'naturalizing' the emerging industrial order:

While urbanism and industrial capitalism were fragmenting an older social unity
(or impression of unity) and replacing it with the more independent alienated
structures of modern society, physiology was presenting holistic images of cohesive
parts in dynamic interaction . . . Regularity yet change, order yet progress; this
was the service of the organismic metaphor.[42]

Such an analysis raises the question of the extent to which the actual
formulation of scientific knowledge – and not just the use to which it
is put – is determined by developments taking place in the wider society.
  One of the most important forms of contemporary contextualism has
links with work being carried out in anthropology. Ever since Durkheim,
social anthropologists have shown a considerable amount of interest in

the relationship between social and natural order – with the knowledge of nature which certain preliterate societies formulate usually being regarded as a reflection of their social organisation. During the 1960s, social anthropologists like Mary Douglas and Robin Horton began to apply this approach to contemporary Western societies.[43] This has had a major impact upon the work of many sociologists of science, particularly those associated with the influential Edinburgh Science Study Unit.[44] Past and present members of this Unit tend to emphasise the role which the scientist's perception of his/her wider social interests plays in the actual formulation of scientific knowledge. This approach is exemplified by Steven Shapin's work on the disputes which arose between phrenologists and anti-phrenologists in nineteenth-century Edinburgh.[45] Although these disputes centred around cerebral anatomy – the precise shape of the cranial bones, the fine structure of the cerebellum, etc. – close observation by the two sides did not lead to a gradual convergence of belief. Shapin accounts for this by arguing that the anatomical knowledge which each side formulated was ultimately designed to further their respective socio-political interest.[46] This leads him to the conclusion that: 'Reality seems capable of sustaining more than one account given of it, depending upon the goals of those who engage with it; and in this instance at least, those goals included considerations in the wider society such as the redistribution of rights and resources among social classes.'[47] When viewed from this 'anthropological' perspective, the scientist becomes 'an active, calculating actor whose intellectual products are crafted to further the variety of his interests.'[48]

It is important to state at this point that none of the leading contextualists is anti-realist, that is, they do not deny that a real world exists independently of the scientists' perceptions and conceptions of it. What they do deny is the belief that scientists possess a method which allows some kind of direct access to reality. To this extent, they are in agreement with contemporary philosophers of science, most of whom point to the limitations imposed by human perceptual and cognitive structures. What is distinctive about contemporary contextualists, however, is the emphasis which they place on the ways in which these structures can be influenced by the scientists' wider social context. It is also important to note that contextualists do not deny science's technical role in society. The Edinburgh School, in particular, stress that scientific knowledge is often multi-functional, in that it may have technical value while at the same time serving a wider socio-political purpose. This ensures that a particular scientific concept or theory cannot be regarded as technically suspect simply because it has a socio-political dimension.

Although only a small proportion of recent studies in the history of science display a clear affiliation with either a neo-Marxist or an anthropological approach, the number of studies which emphasize the scientists' wider social context has increased enormously over the last two decades – a fact which suggests that a broadly contextualist approach may soon come to dominate the discipline.[49] It should be pointed out, however, that there is no clear consensus among social historians of science conerning the relationship between science and the wider society. In fact many of the studies which point to interesting parallels between scientific thought and social structures avoid this issue altogether. Despite this, all contemporary contextualists would agree with the following statement by Barry Barnes and Steven Shapin: 'Increasingly we have become prepared to treat science as an aspect of our culture like any other . . . There is now a real interest in our natural knowledge as a product of our way of life, as something we have constructed rather than something which has been, so to speak, revealed to us.'[50] It is therefore clear that within sociology – as within philosophy – science has increasingly come to be seen as more a product than a discovery. This is a realization towards which the history of science has made a major contribution, but it poses problems to which historians of science must continue to respond.

The Historiographical 'Catch-22'

The intellectual developments outlined above – together with the increasing involvement of philosophers, historians and sociologists in the history of science – have obviously had a considerable influence upon the way in which histories of science are written. By calling into question traditional philosophical and historiographical assumptions they have undermined the traditional consensus among historians of science and encouraged an increasing concern with both the epistemology and the social relations of scientific knowledge. Despite this breakdown of traditional consensus, however, no new consensus has emerged to take its place. In fact it appears that many historians of science are extremely reluctant to confront the full historiographical implications of these recent intellectual developments – hence Shapin's reference to a 'poverty of theory among many historians of science.'[51] Could it be that one of the main reasons for this reluctance is that these developments introduce a fundamental reflexive paradox in the historiography of science? This paradox revolves around the following questions. By calling into question the objectivity of scientific knowledge are not historians of science forced to reflect upon the objectivity of the knowledge which they 'produce'? If no method exists which will allow scientists to arrive at a true knowledge of the natural world, what chance do historians have of constructing a method which will allow them to arrive at the truth about science's past? If scientific knowledge is, in part, a product of the scientist's socio-political context, what basis has the historian for claiming that the production of historical knowledge is independent of his/her socio-political context? This kind of reflexive paradox is not unique to the history of science, but historians of science can no longer avoid these questions, if they are to validate their work.

It is interesting to note Butterfield's response when faced with the problem of formulating an objective historiography, for having criticized the Whig approach, he experienced great difficulty in articulating how true histories should be written. Perhaps this explains why he eventually ended up writing a mild form of Whig history himself – *The Origins of Modern Science* (1949) being a prime example. This raises a second series of questions. Can accounts of the past ever be written other than from the standpoint of the present? Can such accounts have any relevance for contemporary culture unless they are based upon the concerns of the present? If not, what concerns should the history of science address – what should its contemporary role be? It is here that a problem emerges for those who teach the history of science, for it is the way in which the history of science is taught which determines the role which it plays. Its traditional role has been to validate the positivist view of science. This ideology of science may well have addressed the concerns of nineteenth-century Western European society – or, at least, certain sections of that society – but it cannot be regarded as an adequate response to the wider concerns of today. It is those who teach the history of science who must decide which of these concerns the history of science should address. It is they who must decide what its present role should be.

NOTES

1 Figlio (1976).
2 Thackray (1980).

3 Jacob (1973).

4 I have tended to illustrate the main text with examples drawn from histories of 'biology'. This merely reflects the bias of my own reading and it should not be taken to mean that histories of other branches of science are unable to provide similar examples.

5 Delaporte (1982).

6 Mayr (1982).

7 Given his view of the history of science, it is perhaps no surprise to learn that Mayr was once a leading scientist. There is still some disagreement about the extent to which an advanced scientific training is a help or a hindrance when writing histories of science. See, for example, Nathan Reingold's reply to certain comments attributed to Charles G. Gillispie (Reingold 1981). Also see M.A.B. Whitaker's comments on Reingold's paper (Whitaker 1984).

8 Porter (1980).

9 Hall (1983).

10 Heisenberg (1958).

11 Iggers (1975).

12 Bloch and Febvre were not alone in their criticisms of traditional event history. Lamprecht in Germany, Durkheim in France, and Beard in the United States had already denounced the way in which historians tended to concentrate on political and military events. They argued that the past could not be reconstructed by some kind of unbiased immersion in the historical data and pointed out that such data must always be approached with specific questions or hypotheses in mind – questions and hypotheses which would always be determined by the historians' wider social context.

13 Bloch (1949).

14 Bloch and Febvre have had an immense influence upon historiography. The journal they founded in 1929, the *Annales d'histoire économique et sociale*, led to the emergence of the *'Annales'* school which arguably constitutes the single greatest influence upon contemporary historical scholarship. For further details see Aymard (1972) and Stoianovich (1976).

15 Butterfield (1931).

16 Kramer (1986).

17 Ever since the end of the nineteenth century the history of science has been taught within the philosophy departments of French universities. This helps to explain why France has such a long tradition of linking the history and philosophy of science.

18 Bachelard (1934).

19 Hacking (1979).

20 Bhaskar (1975).

21 Despite Bachelard's undoubted importance, there has been a tendency for British and American scholars to neglect his contribution to the history and philosophy of science. His major work *Le Nouvel Esprit Scientifique* (1934) was only translated into English in 1968. The main studies of his work, available in English, are provied by Lecourt (1975), Bhaskar (1975), and Gaukroger (1976).

22 Where Bachelard concentrated on the physical sciences, Canguilhem concentrated almost exclusively on the biological sciences. Few of Canguilhem's

studies have been translated into English but those which have include
Canguilhem (1963), 'The role of analogies and models in biological discovery';
(1978) *On the Normal and the Pathological*; and (1981) 'What is Scientific
Ideology?' The main studies of his work available in English are provided by
Lecourt (1975), Gordon (1980), and Shortland (1981).

23   Quoted in Canguilhem (1970).
24   Canguilhem (1977).
25   One of the best examples of 'precursoritis' is the search for eighteenth century
     precursors to Darwin. This particular search reached its peak around the time
     of the centenary of Darwin's *Origin of Species* in 1959 – see, for example, Eiseley
     (1958), Glass (1959), and Green (1959). For an interesting analysis of the
     function which the notion of the precursor fulfills within histories of science,
     see Sandler (1979).
26   Most of Foucault's major works have been translated into English. Those
     works which are of particular relevance to historians of science include Foucault
     (1970, 1973). It should be pointed out that Foucault was, by training, a
     philosopher and his works – like those of Bachelard and Canguilhem – are
     not exactly an easy read for those with no background in continental philosophy.
     There are, however, a number of useful commentaries available – see, for
     example, Guedon (1977), Sheridan (1980), Weeks (1982), and Jones (1986).
     The introduction to the latter provided the basis for this paper.
27   Foucault (1972).
28   Foucault (1970).
29   Studies which have been influenced by Foucault's archaeological approach
     include Jacob (1973), Albury (1977), and Schaffer (1980). François Jacob's
     work is particularly useful as a highly readable alternative to traditional
     histories of biology. It is also interesting to note that Jacob – like Mayr – was
     originally trained in biology and was awarded the 1965 Nobel prize for
     medicine (see note 7).
30   Crombie (1959).
31   Krige (1980).
32   For two useful reviews of the history of the philosophy of science, see Losee
     (1980) and Oldroyd (1986). The main flaw in these accounts is their almost
     total neglect of Bachelard.
33   Hesse (1980).
34   For an excellent critical account of this recent work in the philosophy of
     science, see Chalmers (1982).
35   Feyerabend (1975).
36   Young (1973).
37   Laudan (1977).
38   See, for example, Jennings (1984).
39   Bernal's Marxist outlook is clearly apparent in his references to the primacy
     of economic structure. For example, in *Science in History* (1954), he says of
     science in the Middle Ages: 'It arose as a consequence of the breakdown of
     the old classical economy and was in turn to decay and vanish with that of
     the feudal economy that succeeded it.'
40   Young (1969).
41   For a brief review of work on the social relations of science which places

particular emphasis on Marxist and neo-Marxist approaches, see Young (1979).

42 Cooter (1979).

43 See, in particular, Douglas (1966) and Horton (1967).

44 Members of this 'Edinburgh school' include Barry Barnes, David Bloor, H.M. Collins, Christopher Lawrence, Donald Mackenzie, and Steven Shapin. For a representative selection of historical studies by some of these individuals, see Barnes (1979).

45 Shapin (1979).

46 This example may reinforce the widespread belief that a sociological approach can only be applied to the 'margins' of science. That this is not the case can be seen from the numerous studies in the history of physics and chemistry which come to broadly similar conclusions to those of Shapin's study. See, for example, Margaret Jacob's studies of Newtonians and anti-newtonians (1976a,b) and James Jacob's studies of Robert Boyle (1977, 1978). It should also be noted that several members of the Edinburgh school have carried out studies of contemporary science – many of which can be found in the journal *Social Studies of Science.*

47 Shapin (1982).

48 Shapin (1980).

49 For two useful reviews of studies in the history of science which emphasize the scientist's wider social context, see Shapin (1980, 1982). One study worthy of special note is Paul Forman's attempt to link the development of quantum mechanics to certain socio-political trends following Germany's defeat in the First World War (Forman 1971; see also Chant 1980). Here Forman argues that post-war criticisms of scientific materialsm and determinism – such as those expressed by Oswald Spengler – encouraged the adoption of indeterminist theories in physics. Because of its subject and its year of publication, Forman's paper is widely regarded as a classic contextualist study.

50 Barnes (1979).

51 Shapin (1979).

Practice

School History, Science, and Technology
An HMI View

Roger Hennessey

One of the main themes in the current debate on the curriculum is that schools ought to prepare young people for life in a 'technology-based society', whether we consider people as citizens, workers or private individuals. History has a clear and central role to play in this enterprise, although as it is presently constituted in many middle and secondary schools it might find the task difficult. If we are to master technology, to bend it to beneficial purposes, and to curb its more sinister aspects, then it is essential to understand its nature and logic. The origins, possibilities and threats of technology can all be illuminated by history.

Definitions

Terms such as 'science', 'technology', and 'history' require working definitions before we can establish any possible or useful relations between them, particularly if we are to consider any consequent pedagogical implications. Technology has been described usefully as 'the application of science to the human condition', and we may call science 'the rational investigation of the universe'. 'History' is a term employed in three distinct ways: the past; a record of the past; and an interpretation of that record – the second variation prevails in English and Welsh schools. Further refinements are possible: 'pure' as opposed to 'applied' science; and 'high', 'low', 'hard' and 'soft' technology. However, a touch of Occam's razor is necessary at this point if the argument is to be contained and coherent. While science and history are disciplines with their own methods and assumptions, technology is much more interdisciplinary; it cannot be comprehended by science, history or craft alone.

History, science, and technology can be combined or related in four distinct ways. First, it has been claimed that history is a science itself: J.B. Bury made this assertion in 1902, in his inaugural address as Regius Professor of History at the University of Cambridge. However, although history can and should employ many of the elements of true scientific method (observation, selection, creating and testing hypotheses), there are some important differences, notably that historians cannot test their hypotheses in a series of controlled experiments. A second connection is that of history and science helping each other through the medium of technology, for example in dating techniques (such as 'dendrochronology', i.e. tree-ring dating, or archaeo-metallurgy); in return, historians' familiarity with long runs of archives and diaries has aided the work of climatologists. Thirdly, there are the histories of science and of technology. These are rather specialized areas, perhaps too narrow for general school use, although there has been at least one GCE 'history of science' syllabus, and some of the current AS Level proposals lend themselves to this theme. The fourth area concerns us most: the study of the effects of science and technology on human development. Here lies the largest set of usable possibilities for teachers of history and related areas of the curriculum.

Public examination returns and tentative generalizations from HMI visits to schools suggest that most pupils study some history between 9 and 14, and that about half opt for it as a specialism between 14 and 16. About half of the specialists take traditional 'political/diplomatic' history and about one-third take economic or social-and-economic history: the distinction is important as will be apparent later. For the most part, non-examination 14 to 16-year-old pupils incline to social and economic history. In 1985, there were about 290 000 candidates for the varieties of 16+ history. But about half of school pupils drop history at 14; in all probability they will have had a 'chronological run' up to that point, or they will have been exposed to a series of topics.

There is already plenty of technological history scattered around in this national history pattern, but in a very haphazard and unsystematic way. It cannot be said that this arrangement does very much to explain the nature of technology and its historic relationship with social development. Typical and time-hallowed technological items in 'standard' history are: early hunting implements; the pyramids; bits and pieces of military engineering; some early industrial processes (usually concerned with the textile industries); canals and railways. Frequently these artefacts are taken as read and are not treated as symbols of something underlying human history. The flowering of late-nineteenth-century technology has traditionally received scant attention; consequently the telephone, cinema, radio, gramophone, aeroplane or motor car are placed lower in the hierarchy of historical esteem than that familiar textbook icon, the spinning jenny. The main charges against this development of technological content in history courses are that it lacks system; it does not attempt to explain or analyse technology as such, it is arbitrary in its choice of evidence; and, above all, it is tacked on to another framework of reference (political history in the main) rather than attempting to treat technology as a foundation of civilized life.

Common Perception

The main reason for this common perception of history as a political chronology can itself be explained historically. Much past history was written about, and to a large extent by and for, the governing and political classes. Technology was not of major interest to these people, although clearly they made use of it like anyone else. This tradition of political history received its first main challenge from social historians like J.R. Green in the nineteenth century. They wanted their history, as Thackeray had it, 'more familiar than heroic' and tended to share Voltaire's feeling about the aridity of history as a mere record of 'one barbarian king succeeding another on the banks of the Oxus or the Jaxartes' – or, for that matter, one cabinet succeeding another on the banks of the Thames. Although at first there was a tendency to treat 'industry' or 'the poor' as problems to be solved or mitigated by political means, the development was crucial. It helped to give rise to the rich diversity of history now on offer both in academic and popular circles: industrial archaeology, with its flourishing local associations; business history; military history; demographic, urban, econometric and psycho-history. Many of these variations are concerned with technology. The heart of the matter is exemplified by the motor manufacturer, Henry Ford, who said: 'Records of old wars mean nothing to me – history is more or less bunk, it's tradition'. On the other hand, when setting up his Greenfield village museum he stated 'I'm going to give the people an idea of real history . . . we are going to show just what happened in years gone by' – precisely the aim of Leopold von Ranke, a leading nineteenth-century historian who thought that history ought to show

'how it actually happened'. The one worked in archives, the other assembled domestic and industrial gadgetry; both served history well if rather one-sidedly in their respective roles. The emancipation of history has gone a long way beyond higher education: the shelves of national booksellers are clear evidence of this, crowded with books on old buildings, motor cars, railways, warfare, and the dresses, dishes and entertainments of our ancestors. Museums, television, and the tourist industry have all gauged this major change quite accurately.

In turn, history in schools has been suffused with the new approach to the discipline, although an informal HMI survey of GCE syllabuses and papers some 10 years ago found that the incidence of technological elements (generously interpreted) was only 4% in 'political' papers but 25% in social and economic ones. A harder interpretation reduced the incidence in both cases and it was so slight in the political papers that it barely registered. A recent updating scrutiny found the position much the same, with possibly a very slight improvement. A clear ray of light is the popular topic 'Medicine through Time' taken as part of the Schools History Project course (leading to an examination at 16+). This topic has now been joined by a similar one on energy, and both are discussed by Joe Scott later in this book.

The generally low incidence of science and technology in 'orthodox' history syllabuses will not encourage teachers and pupils to depart very far from mainline political or social history. Further evidence for the growing interest in technological history is in the high number of 11 to 16-year-olds who reply to oral questions about change, progress, the price of change and so on, in technological terms (for example the convenience of household machinery, the threat of nuclear warfare, the relief of suffering afforded by modern drugs). Examples cited by 'political' historians are usually 'off the syllabus'. Many history teachers, of course, have a time-honoured tradition of employing anecdotes and material outside the syllabus to broaden their courses. The answers did not seem to be significantly different for any ability level.

The question nevertheless remains: why should we bother to inject science and technology into our history courses? In turn, if we do this, how are we to fit such elements into crowded syllabuses? We should add science and technology to our history because they have been among the principal foundations of civilization, and powerful forces behind social and political change. It is a contemporary and occidental conceit to think that technology is modern and Western. Thomas Carlyle praised the timeless importance of 'the nameless boor who first hammered out an iron spade'; he certainly applied science most effectively to the human condition. The dazzling achievements of oriental and Islamic science and technology are a part of our common heritage. We respect the masons of the Acropolis or Chartres cathedral as we respect the masters of electronics today – for their professional style and technique. This leads to another reason for including these things in history courses: science and technology have been common experiences, at one level or another, of the whole human family. Teachers may wish to draw attention to this 'unity in diversity' of our many cultures. Any 'world' or 'European' studies courses that omit this consideration are very partial and imperfect, and have missed a potent opportunity. We are reminded by the White Paper, *Better Schools* (Cmnd 9469, HMSO, London, 1985), that one of the main purposes of learning at school is to help pupils to appreciate 'human achievement and aspirations'. This assertion is further supported by a recent HMI publication, *History in the Primary and Secondary Years: an HMI View* (HMSO, London 1985), which states, *inter alia*, that a history syllabus ought to help young people learn about 'the effect on our lives . . . of science and technology'. Science and technology, like the arts, are areas where human achievement has been varied and inspiring, as well as being of everyday utility.

Difficulties and Opportunities

And yet even a slight acquaintance with technology reveals that it generates difficulties as well as opportunities. One of the dangers in advocating technology on a syllabus is that the optimum will be confused with the maximum, as is very often the case when new causes are pleaded in education. This can only lead to an uncritical acceptance of coarse materialism or of an inexorable 'march of progress'. This route leads to the world of the techno-cranks, like the 'Steam Intellect Society' lampooned by Thomas Love Peacock in *Crochet Castle*, or represented by the sinister Bazarov (Turgenev's *Fathers and Children*), who thought 'a decent chemist is worthy twenty poets', an assertion that begs many questions, to say the least.

It is untrue that history is little more than the story of the generation and utilization of kilowatt-hours, and any school of history that subscribes to such doctrines will not prepare young people for society as it is, or as most sane people would wish it to be. There is, however, an equally pernicious tradition (also dating from the industrial revolution) that asserts that technology is unnatural and that it enslaves us. This prejudice is not without influence in our schools. It is based on historical ignorance and maintains itself by filtering out any consideration of the very liberating nature of so many familiar technical devices: glass windows, which freed our forebears from darkness and draughts; bicycles, which opened out horizons for city dwellers a century ago; typewriters, which arguably did as much for women's liberation as any extension of the franchise; and, of course, surgery and pharmaceuticals, which have freed so many people from fear and suffering. This prejudice also ignores the manifold ways in which technology has extended the scope of the arts and has presented the results to large and popular audiences. The literary and humanistic suspicion of technology is old and influential and it has played a part in over-balancing the school curriculum against its enemy. It was a symbolic, if sad, irony that an early high-priest of this doctrine, Matthew Arnold, met his death while running to catch a Liverpool tram.

The Right Balance

Curriculum-planners have to get this balance right according to their own lights. The application of science to the human condition has many implications: moral, ethical, and political, as well as economic and social. Young people need to be guided carefully in this area, not subjected to unexamined folklore, and history can provide some useful data for this exercise.

For those teachers who wish to adapt history courses to admit science and technology according to a rational plan, there are snares to avoid and organizational problems to face. First, it is important to avoid the heroic tendencies of history, thinking solely in terms of great inventors or scientists, thus exchanging one 'grand history' for another. This kind of approach misleads, for while it is true that the individual efforts of people like Edison, Marie Curie, or St Benezet (the only canonized engineer to date) are most worthy of record, yet even Newton thought he only saw as far as he did because he 'stood on the shoulders of giants'. Science and technology are great collective efforts involving millions of people. Also, technology is just as evident in a bathroom tap or a rope suspension bridge as it is in a man-made satellite or in papers read to learned societies, and it is immanent in much of our social and economic life. To advocate more science and technology in history courses is not, emphatically, to advocate a great intellectual revolution. It is to suggest rather more of the engineers and artisans and rather less of the chancellors and kings. This is a question of adjustment, not of massive inundation. Anyone suggesting or implementing these changes will have to face problems implied logically by this article. If school history is to relate more precisely to our competence in dealing with the problems of a high-technology society, then we have to ask questions about: fashioning an overall school curriculum that will encourage or require history to illuminate the problems of technology and its effects upon our civilization; the appropriate initial and in-service training for teachers; the production of appropriate resources; the requirements of external examinations and their syllabuses; and the relations between history and other disciplines. Already many history teachers and a few courses (the Schools History Project, for example) have shown what can be done within present constraints, so it follows that an initial and cost-effective strategy would be to rationalize that which is, before turning to that which ought to be.

Very often we fail to see the obvious because we are immersed in it and have never 'stepped back'. Is it not curious, however, that a discipline we call 'history' has many courses that barely touch upon our highest technical and cultural achievements? There are 'history' syllabuses covering the nineteenth century that never mention Beethoven, Gauss, Brunel, telegraphs or tinned food, but pay assiduous attention to Gladstone's second cabinet or the treaty of Unkiar-Skelessi. There are students of the 'seventeenth century' who have never heard of Kepler or the baroque. It is arguable that Darwin and Copernicus radically affected the ways in which we perceive our condition, but they are accorded less history syllabus time than Charlemagne or Bismarck: once again, the balance is wrong. Unless history can show a greater awareness of these matters, it may be open to charges of bias even of irrelevance. While history may not be twenty times as valuable as chemistry (with deference to Bazarov), then at least it might demonstrate that it was no less valuable.

Postscript, December 1988

A great deal has occurred since this paper was delivered at Oxford over a year ago. Private initiatives have brought about some encouraging developments, such as the 'Energy through time' unit of the Schools History Project, alluded to above and described elsewhere in this book by Joe Scott; and that element of Advanced-level history being pioneered by the Cambridge History Project, which is concerned with a long-term study of industrial and technological development.

The development of 'coursework' in GCSE history has not infrequently been linked with themes related to industrial history, often touching on technological issues. The potential for the history of science and technology would seem to be considerable in this field. The SEAC (and it predecessor, the SEC) have urged the designers of Advanced-level history courses to take a broader view of the subject, and not to deny the history of science and technology their places in the sun.

There are two other public initiatives which are full of potential. HM Inspectors have written a document (*History from 5 to 16*, HMSO, 1988) which affirms the importance of historical 'content' and which gives strong support to the history of science and technology as one means by which we may broaden the meaning of 'history'. This publication was welcomed by the Secretary of State for Education and Science as a 'valuable contribution' to the debate on the National Curriculum. The National Curriculum is a major initiative which has accorded history the status of a 'foundation subject'. It will have to be studied by all children from 5 to 16. Anyone who is anxious that the history of science, or of technology, should be part of this programme can make their views known to a history working group if and when one is established. In short, our times are highly propitious for people who seek to encourage the history of science and technology – providing they seize the opportunities being offered.

Medicine and Energy in the SHP History Syllabus

Joe Scott

The traditional school history syllabus has for a long time included some aspects of technological and scientific history. The Middle Ages would make little sense without some understanding of the agricultural technology of ox-drawn ploughs and crop rotations. Every schoolboy or girl has heard of the two 'master tools' of the new Europe that emerged at the end of the Middle Ages – the three masted ship and the technique of printing. The scientific revolution that began at about the same time has always figured at least at the 'trivial pursuits' level of an honourable mention in the school text-books for Copernicus, Galileo, Newton, and the Royal Society. The steam engine, the new textile technologies, and the railways of the early nineteenth century were clearly the motive forces of massive social and economic changes, and the school history syllabus has given them due weight almost since the days of Samuel Smiles.

But school history tends to be made up of a series of separate episodes – this week George Stephenson, next week the Reform Act of 1832. School students of history are likely to know very little about the scientific and technological background out of which a development emerged. A history teacher teaching about Newcomen's steam pump, for instance, cannot assume any knowledge about how new the idea of atmospheric pressure was in the seventeenth century, and may not have time to explain about von Geuricke or Papin. So Newcomen appears rather inexplicably with his atmospheric engine, just as a few terms earlier Henry the Eighth appeared inexplicably with his six wives. Children in any case tend to latch on far more easily to episodes and personalities than to general ideas and intellectual trends. So perhaps what they finish up with at the end of their study of history is at best the conviction that scientific and technological changes have played an important and interesting part in history, especially in the last three hundred years. They may also be able to describe some of these changes, especially the technological ones, in a rather superficial way. Some of them may also have learned a more harmful lesson – that the scientific background and finer technical details are inexplicable matters best left to experts in brown or white coats who are at home in the alien culture of the science lab or the workshop.

The Schools History Project

The Schools History Project (SHP, formerly Schools Council History Project) did not set out to question the content of the history syllabus or to argue whether it should include more or less of the history of technology and science. It aimed to re-state the aims of the syllabus in terms of the needs of adolescents to develop their skills and their grasp of concepts. What needs was the discipline of history uniquely qualified to meet? The Project Team identified five such 'adolescent needs', and designed a syllabus with four components, each targetted on one specific need but all in common aiming at a central target, the fifth 'need' – *to develop the ability to think critically about evidence and to make judgements about human affairs*.

One of the four other 'needs' which the Project saw as important was the need to understand change and continuity, and in order to meet this a 'Study in Development' was to be a key element in the course, taking about one and a half terms in a two year examination course for O-level and CSE (now GCSE). The purpose of this study in development is not to teach pupils about a particular development chosen as being of key importance to them. Even less is it to enable them to draw some general philosophic conclusions about change and development in history. It is to provide them with some experience in thinking about change, and development, and the factors which inhibited or encouraged them, so that they can use terms like development, cause, factor, progress, regress, turning point, etc., and have some understanding in one context at any rate of what we mean by change and development in history.

The topic for the Study in Development which the Project Team chose in 1973 was *Medicine Through Time*, and they produced a set of teaching materials under that title. The syllabus has been in use since 1974, and has clearly been very successful if measured in numerical terms. In 1987 about 50 000 sixteen-year-olds sat Schools History Project examination papers including the section on the history of medicine. This is about one third of all those taking history examinations of any sort at 16+. All the GCSE examining groups have a similar syllabus on offer for 1988 and beyond, and it looks as if the coming of GCSE will see a further rise in number taking the SHP syllabus.

The Project Team always intended to suggest alternative development studies, but it was not until 1982 that the topic *Energy Through Time* was chosen. This second technological choice was deliberately made with the fashionable idea of contributing to an increase in technological literacy. It was made after consultation with teachers, though there were of course strong lobbies for other topics. If and when a third alternative appears it is likely to be a sociological, one like *Crime and Punishment or Education*. The *Energy Through Time* materials were published in 1986, so only a tiny handful of candidates sat papers on this topic in 1987, although there will be several thousand in 1988. It remains to be seen how successful it works, though the early indications are encouraging.

Perhaps I should explain my own role in these developments. I was first a teacher in a 'trial school' for the Project materials and then for some years National Coordinator of the Project. Since my retirement in 1982 I have written the *Energy Through Time* materials and the new version of *Medical Through Time*. I may say that I personally very much welcomed the technical and scientific topics. I am myself a casualty of our 'two cultures' educational system, though I hope a reasonably technically and scientifically minded one. At the tender age of thirteen I was persuaded by a classical headmaster that I should take up Greek. This meant in effect no more science. As far as technology was concerned I had already made a table lamp with a mortise and tenon joint so what more was there to learn? As a result I have had to cope with the world of nuclear bombs, the big bang theory, the double helix, heart transplants, hi-fi, computers, and do-it-yourself, on the basis of School Certificate Greek. Things have improved a little since the 1930s but it has always seemed to me as a teacher that we create arts/science subject-mindedness by our narrow options system. Anything we can do within the history syllabus to minimize this would do good. It was for these reasons that I welcomed medicine as a topic and seized the chance to work on the energy materials.

Results

What can I report from the Schools History Project's experience that might interest readers of this book?

The first point is that the difficulties have not been due to the technical or scientific problems. History teachers dealing, for instance, with Harvey's work on the circulation have sometimes welcomed the help and cooperation of colleagues from the biology department, but they have usually managed perfectly well on their own. It is not very difficult to follow the outline story of the establishment of the theory of the circulation of the blood from Vesalius' insistence that there were no holes in the septum, through Harvey's triumphant and exact demonstrations, to the final vindication when Malpighi was able to see the capillaries. It is more difficult to get pupils to see why many people at the time opposed the new theory, or why it had so little effect on medical practice for so long.

It is more difficult to get pupils to enter into a discarded set of scientific ideas than into an accepted one. They tend, for instance to be very familiar with the idea of germs, and to make a pretty clear distinction between living and dead matter, so that that idea of spontaneous generation seems a silly one. This is no basis on which to understand Pasteur's achievement. But problems of this sort are ones with which we as history teachers are familiar. A key feature of our task after all is to enable pupils to enter into the patterns of thought of people of a different time and place. It is by exercising the historian's ability to empathize that pupils can come to see how unlikely the germ theory was in a world without microscopes, just as they can see how unlikely the heliocentric theory was in a world without telescopes.

As an example of the successful recreation of a discarded set of ideas, here is the answer of an able fourth year pupil to the question: 'Explain the Greek theory of the four humours. Why did it continue to be accepted for so long?'

Hippocrates was a Greek doctor and scientific thinker. The Greek scientists believed that everything was made out of four main elements, earth, air, water, and fire, and in the human body these were linked to four liquids or humours. They also believed in things being properly balanced. If the elements were out of balance there was a storm or an earthquake. If a man's humours were out of balance he was sick. The obvious thing to do was to restore the balance, by drawing off excess of blood or by giving honey to bring up phlegm. Some of these cures worked and some did harm, but as long as men believed in the Greek ideas of chemistry, about the elements, and of physics about the need for balance, the theory of the four humours fitted in very well. So people went on believing it until the new ideas of physics and chemistry came in in the eighteenth century.

The boy who wrote this succeeds, I think, in showing why the theory of the four humours made sense when viewed as part of the Greek pattern of thought.

The greatest difficulty that teachers found with this course was in giving pupils a usable overview of a very long and complex set of changes. The answer I have just quoted shows that this is attainable. The writer clearly had in his mind a picture of a Greek scientific world view which remained acceptable until a new science emerged at a time which he rather vaguely characterizes as 'the eighteenth century'.

This problem of overview is one which faces all teachers of history and perhaps of any other complex subject, but it is specially important in a 'Study in Development', which by taking development as its subject matter absolutely requires an overview of the thing which developed. Project teachers have tackled this problem with some success by adopting the following strategy. First a rapid outline narrative of the whole development is built up, with a heavy emphasis on turning points which give shape to the story as a whole. Then one or two individual 'factors' are studied in detail – 'How has war influenced medical change?' Finally specific problems are tackled 'in the round': 'Why did this important change (e.g. the coming of penicillin) happen when and where it did?' If we imagine this strategy applied to the coming of the steam engine, the first overview narrative need not worry about von Guericke or Joseph Black, but will briefly cover the work of Newcomen and Watt, and try also to bring out why this new prime mover was important in relation to the industrial revolution and to our modern industrial society. Then later, when pupils have gained some overview and are studying how science influenced developments in man's use of energy, they will return to the seventeenth and eighteenth centuries with time to put scientific ideas about the vacuum or about the measurement of heat in the centre of the focus. Finally towards the end of the course, after dealing with several such factors, they may come back to the question 'Why did the Steam Engine develop in eighteenth-century Britain?'. They may then be able to put together the scientific ideas, the need for new sources of power in the mines and elsewhere, the coming of new metal-working techniques like those of Darby or Wilkinson, and the attitude of enterprising capitalists like Boulton, into a convincing answer to the question. This ability to see and to begin to understand the web of causative factors – scientific, technological, economic, social, and political – that actually made things happen is (or should be if it isn't) the real aim both of teachers of history and of historians of science and technology.

One danger of studying a technological or scientific line of development can be illustrated by a passage from a candidate's answer in a *Medicine Through Time* examination. 'Then, as the times in history got more modern so did the methods or cures, until eventually the same cure was being done but they bettered it every time, so it is now coming to us perfect, where nothing can go wrong.' A history of medicine or of energy or anything else can seem to be the story of a series of cumulative scientific or technical discoveries culminating in 'clever us'. The danger is more serious since most children are likely to start the course holding this naive idea of a sort of inevitable progress. Perhaps the history of medicine is particularly prone to this problem in view of the enormous achievements of high technology medicine in this century, so it is important to include thalidomide as well as penicillin, and acupuncture as well as anaesthetics. Perhaps in the history of energy the 'clever us' syndrome is likely to be less of a problem in the post-Chernobyl decades, as various doomsday scenarios work their way into popular consciousness. It could be indeed that there may be an opposite problem. The children of late twentieth-century Guardian readers may allow their awareness of pollution, the limited world reserves of fossil fuels and the like, to distort their understanding of the exciting achievements of people like James Watt or Ernest Rutherford.

I think, however, that the best answer to both these dangers is for the history teacher to do what he or she may in any case look on as an essential part of their ordinary task – to encourage pupils to understand and respect the attitudes and ideas of people in the past, and to try to make sense of them in their own terms, even though they seem strange and alien to late twentieth century eyes. Perhaps I could finish with the response from a CSE examination a few years ago that shows one sixteen-year-old doing exactly that. The question was: 'Why did physicians continue to use and patients to pay for medical treatments that were useless or even did more harm than good?' Part of the answer ran:

Physicians didn't mind when the incantations and potions didn't work because it was like saying your prayers and sacrifices to the gods, if the gods didn't answer your prayers it didn't mean the prayer was said improperly it just meant the gods were not helping you. It was the same with medicine, all their ideas were magical and mystical, and so the physician had to do it right like the priests, yet it was up to the gods in the end. So when they didn't get better it was like the gods saying 'No you can't have any gold and dancing girls today' and this didn't stop you saying your prayers and sacrificing heifers.

History and Philosophy of Science and Multicultural Science Teaching

Piyo Rattansi

Introduction

Can school-science contribute to the aim of fostering a 'multicultural' perspective?

It is not evident at first sight how school-science, as such, can have any influence, one way or the other, on multiculturalism. In other subjects it may be possible to pinpoint ways of teaching which may buttress a negative image of the parental or ancestral cultures of non-European pupils, thereby helping to sustain racist stereotypes and racism, and to suggest how such a bias could be mitigated or eliminated. Science, however, is taught through the teaching of principles and theories, supported by practical and laboratory work. It is difficult to see how school-science teaching can either help or hinder the aim of fostering of a multicultural perspective.

That may, however, be too simple a view. A case can be made out for suggesting that current school-science teaching, when examined in the light of the history and philosophy of science, emerges as promoting, however unintentionally, a deeply negative image of non-Western societies and cultures. The dominant place of science as a feature of twentieth century life and culture makes that a fact of great importance for the self-image and sense of identity of pupils from the ethnic minorities.

School Science Teaching Today

One of the more progressive features of school-science teaching currently, as compared with the methods which generally prevailed about thirty years ago, is the 'discovery' method. In this country it is associated with the 'Nuffield' approach, and in the USA with the integrated courses for teaching Physics and Biology that followed in the wake of the Russian Sputnik. There can be little doubt that it has been a change for the better. Science was turned into a far more exciting study for all pupils when they were invited to enter into a process of rediscovery, rather than expected to master facts by note-taking and reading. Observing, experimenting, and discovering regularities which can be stated in a quantitative form: pupils were encouraged to engage in those activities from the very beginning in their studies of science. They undoubtedly gained a confidence in their own ability to explore the natural world which the old ways of teaching may never have imparted to most pupils.

It is easy to see the advantages of the new teaching over the old methods and approach. But to the historian and philosopher of science, the new methods may seem to have brought their own dangers. These are of considerable cultural significance, since they concern the cultural divide between the arts and sciences in modern societies. But they may also contribute to the formation of attitudes to non-Western cultures (and, indeed, to the pre-modern European past) which are relevant to multiculturalism.

Perhaps the best way to illustrate that is by taking a few short sections from a representative school text and try to show what may at first sight seem unduly alarmist criticisms. They are from *Physics 11 – 13* by John L. Lewis,[1] intended according to the preface as a basic course in physics for pupils aged 11–13, which 'uses a modern experimental approach to impart an *understanding* of the principles of physics.' The book begins with a chapter on 'Materials', which divides up the variety of materials we encounter into the three phases of matter:

In our homes, in our schools, in the streets and in the countryside we are surrounded by all kinds of things.

There is also an immense variety in the materials of which these things are made. In your home there are tables and chairs made of wood, saucepans of aluminium, windows of glass, curtains of cotton, bowls of polythene, bicycle frames of iron, electrical wires of copper.

In your laboratory you will have seen a variety of different kinds of materials. Some of the substances are *solid*: iron, copper, aluminium, rubber, brick, lead, wood, polythene . . . These substances have very varied properties. Some are hard, some soft . . . they may differ in smell, in colour and in their feel or texture . . .

Some of the substances you have seen at home or in your laboratory are *liquid*: water, paraffin, mercury, treacle, vinegar. They too have varied properties: some

for example flow very easily, others are very sticky, some smell, some are coloured.
   Yet other substances are *gases*, like the air around us. Perhaps you have seen balloons filled with different kinds of gases . . .

The chapter which follows is one on crystals and suggests experiments to grow hypo, alum, salt, and salt crystals. Experiments with polystyrene spheres arranged in a hexagonal pattern are introduced as a plausible model of the underlying structure of crystals, leading finally to the conclusion:

All the evidence you have seen suggests that a reasonable way to make a solid is to use building blocks of small particles. We might suppose the particles are atoms, and we will assume that they are until we get any evidence to the contrary. Of course we do not know why the particles hold together: we shall have to wait until a later stage before we find out.

What has been accomplished so far? The pupil has been led to reflect on the three forms in which matter is to be found around him. He or she has then been introduced to a model, a particulate or atomic one, which can make sense of all these forms and their characteristics in terms of one universal underlying matter with its associated forces. How can anyone possibly object to this way of gradually initiating pupils into modern science? The short answer is that it systematically deflects attention from the enormous array of assumptions which have been smuggled in, which determine not only the 'observations' which seem to be relevant and suggest the experiments to be performed, but the inferences which may legitimately be drawn from them. It gives the pupil the quite misleading impression that it is possible to pass almost automatically from 'observations' to scientific conclusions. By implication, therefore, it must raise puzzles about the obvious incapacity of cultures in the past and in non-Western societies to make such discoveries by using the 'scientific method' which even a thirteen-year-old can deploy today with such ease.

Physics *c.* 1650

Let me try to indicate some of those assumptions by showing how *Physics 11 – 13* could be rewritten to conform to the textbooks on physics which were still being used at Cambridge when Newton was an undergraduate there in the 1660s. (For this purpose I shall make free use of the particular scholastic textbooks from which Newton himself took notes.[2])

'Of the bodies with which we are surrounded, some are solid, e.g. a stone; or liquid, e.g. water; aeriform, e.g. steam or smoke, or radiant, fire or sun and stars. Are they all made of something more fundamental, and, if so, is whatever underlies them one or multiple?

In answering such questions, it is inappropriate to rely on authority. We must attain a knowledge of the *reasons* on which the facts known to us through our senses depend.

Suppose we try to burn portions of some of the solid bodies we encounter. Coal, wood, and paper will catch fire and give off a great quantity of smoke. Metals will not; but they can be melted and turn into vapour when heated for a very long time, some more easily than others. Liquids can readily be turned into vapour, e.g. water into steam. Again, liquids assume a solid form when frozen: water in winter turns into ice, which is hard as stone.

All the bodies of which we have experience on earth, then, assume one of four forms. We may explain the variety of forms by assuming that underlying each is a different element or principle. These may be called earth, water, air, and fire. But since they can be transformed into each other, each body is really a *mixture* of elements, with one or more predominating to give it its characteristic *complexion*.

Bodies gain solidity from the principle we shall call 'Earth'. They may assume liquid form if they contain an excess of 'Water'. Or they may be aeriform if constituted mostly of 'Air'. There is another principle which enters into the composition of all bodies, 'Fire'. Whether they readily catch fire or not will depend on the scarcity or abundance of 'Fire' within them. Similarly all solid bodies contain 'Water', but not all will readily turn into a liquid when heated even for a long time.

Each of the elements has a pair of associated qualities, the ones that are most accessible to us through the sense of touch. They may be warm or cold, wet or dry. Water is wet and cold, fire dry and warm or hot. When we heat water (cold and wet), we are replacing its coldness by hotness, until we have succeeded in turning it into steam or air (hot and wet). (Try to explain water – cold and wet – turning into ice – cold and dry.)

The viewpoint we have adopted depends on the assumption that there is a plurality of elements, four in numbers. But would it not be simpler instead to suppose that everything is made of only one kind of matter? Greek thinkers who flourished at the time of Socrates suggested that all matter could ultimately be resolved into atoms, subsensible 'uncuttable' particles, which differed from one another only in shape, size, motion, or in mutual arrangement. All the variety of qualities – taste, texture, odour, colour, hotness or coldness – would arise from nothing more than the way these primary features of the particles affect our senses. Triangular particles, for example, would cut the tongue and produce the sensation of acidity, while smoother and spherical particles would convey a sensation of sweetness.

You merely have to consult your daily experience of natural processes to realize that atomic explanations are utterly unsatisfactory and must be false:

1. Organized entities cannot be reduced merely to a juxtaposition of their parts. A house is more than a collection of bricks, a word more than a jumble of alphabetical characters: the first has a plan which the architect has imposed on it, the other a meaning. Thus each possesses a *form* in addition to its underlying *matter*, making it what it is.

2. If a mechanical juggling of parts was all that was required if a body made predominantly of one element was to be turned into one of another sort, how could we account for the fact that freezing must turn a liquid into a solid, and heating will turn a liquid into air? There is a regular sequence involved in such transformations, very different from the randomness we should expect from a mere juggling of parts.

3. Atomism reduces all change to change of size, shape, or mutual relation of parts, that is, to a change in the quantitative features of bodies. That is to confuse two different kinds of science, with different kinds of subject-domains. Quantities belong to the domain of Mathematics. But the kinds of quantities with which Mathematics deals are ideal entities, such as points without dimension or lines without breadth. The mathematician derives them by taking the limits of actually existing bodies and thinking away the matter. The only changes in his ideal realm of geometrical bodies are quantitative ones. But the world in which we live and to which our senses give us access, the world of Physics, is a messy and imprecise one of irreducibly *qualitative* change. Aristotle often used a simple example to illustrate that. It was that of Socrates, whom nature had given a snub nose. A mathematician could abstract from that nose a concave curve and discuss its mathematical properties. Would that really contribute anything very important to our understanding of it? For its scientific understanding, the shape of a nose is significant only in so far as it contributes to its functioning as a nose. The man of science will rather use the word 'snub' to describe it, for the idea of a curved shape, and its inhering in a particular sort of biological entity, are incapsulated in that word. Now the entities which actually exist in the world, as it is studied by science, are more like 'snub' than like 'concave'. Therefore the basic principles used to explain changes in it must be of the sort that are themselves capable of change. That is why the four elements, and their associated four qualities, are far more useful in understanding the bodies around us and making sense of the changes they undergo than any principles drawn from Mathematics, the science of the immobile and the insensible. What could be more absurd than to use mathematical entities to explain the subjects of Physics, of the science of the immobile and insensible to explain those of the science of the mobile and the sensible?

It is regularity, fixity, and stability which, as daily experience amply confirms, characterizes all change in the realm of plants, animals, and human beings. That is what makes the atomistic picture of insensible particles blindly coming together or separating, and giving rise to all that we observe in the world, so wildly implausible. How can order come spontaneously from disorder, or regularity from chaos? But we must not assume that regularity is a mark solely of the realm of organic change. It is no less absent from that of bodies which do not appear to contain a living principle in them. Acorns, we know, always develop into oaks, babies into adults, if the natural process of growth is not interfered with. We should ponder on the fact that stones always fall downwards, while smoke and fire always rise upwards, unless forced to do the contrary. Their behaviour, too, seems directed towards a particular and specific aim, as if towards a natural place in which alone they can rest. That must be because each element tends invariably towards the place that, in a rationally-arranged world, it ought to occupy, depending on whether it is heavy or light. The heaviest (earth) would then find rest when it is at the centre of the universe while the lightest (fire) the outermost position, with air and water occupying the intermediate regions. Indeed, force has to be applied to such bodies to make them tend away from their 'natural' places. That, incidentally, reinforces what sense-experience teaches: that the earth occupies the centre of the world, with fire at the periphery of that part of the world subject to incessant change, just below the sphere carrying the moon, and those of the other celestial bodies which whirl around it. Beyond the moon lies a realm in which the only change is one of position and not of substance, a change which can be specified in timeless mathematical terms quite different from the irreducibly qualitative ones with which of a science of Physics must concern itself . . .'.

'Facts' and 'Conclusions'

It is hoped that the juxtaposition of a modern school-science with scholastic-science as it existed in Newton's youth that has been presented above makes it easier to grasp how the same commonly available facts led Aristotle and his followers in succeeding centuries to arrive at conclusions so startlingly different from those which we seem almost 'naturally' to derive from them. They seem 'natural' only because of the background of assumptions that we take for granted, no less than they did. That background constitutes an 'ontology': a picture of the entities which fundamentally compose the world, and, therefore, of the sorts of methods appropriate to its study, and of the kinds of explanations which can appear satisfactory.

The Aristotelian world was qualitative, hierarchically split between celestial and terrestrial, and ruled by purposes immanent in natural bodies, impelling them towards the form of their kind or species. That of current science is the one which began to replace it in the seventeenth century: a quantitative, much more unified one, of change as ultimately due to bits of matter acting on other bits of matter through impact or pressure or through various forces associated with them.

What tends to be blocked out in the current method of presenting school science is the 'traditional' aspect of science: the fact that in science as in any other discipline the individual is initiated into an inherited framework of facts and assumptions which he has to master. Instead, pupils are encouraged to regard themselves as setting out with minds unsullied by presuppositions and hypotheses, armed only with a 'scientific method' which will enable them to process observed facts into regularities and quantitative laws. It must thereby raise grave puzzles about the failure of people in the past, or in other cultures, to make the same discoveries as the pupil today just beginning his study of science. It invites such explanations as the one that their minds perhaps were not as 'well-developed' as our own, or that they were prevented from functioning normally by the fetters of ignorance, prejudice, and superstition. It also sets science quite apart from other intellectual disciplines, breeding an attitude of superiority and disdain for them among its practitioners, which in turn may evoke a 'humanistic' backlash which is just as damaging and impoverishing to our cultural well-being.

By omitting the conceptual framework within which the scientific enterprise must function, current school-science textbooks are apt to leave pupils profoundly unaware of the social and communal character of science. Pitted against the great corporate structures of the medieval world, early modern science linked itself with the emerging individualism of a new epoch, and we have inherited from it its ideal self-image of the man of science as a lonely pioneer (Newton and the apple or Galileo watching the swing of a lamp in church, dropping cannon-balls from the tower of Pisa, or turning his telescope to the heavens), challenging authority and tradition only to pay with years of neglect, derision, or downright persecution. We are thereby likely to dismiss Newton's famous declaration as nothing but the modesty of genius: 'If I have seen further it is by standing on the shoulders of giants.'

Those 'giants', as we know from historical studies, stand for the inheritance of many cultures over millennia: of Sumer, Babylon, and Egypt; of India and China; of the Hellenic and Hellenistic worlds; of the Semitic culture in which Judaism, Christianity, and Islam originated, and of the civilizations they shaped; of the European Renaissance and Reformation – all contributing in complex ways to the birth of modern science in the sixteenth and seventeenth centures.

Conclusion

1 Teaching science through discovery may, however unwittingly, inculcate a deeply misleading conception of the scientific enterprise with profoundly damaging cultural consequences, and in particular

a devaluation of the past and of non-Western cultures.

2   'Observations' do not lead inevitably to 'conclusions' in the way which is made to seem all too natural in school-science textbooks, except against the background of assumptions which constitute a conceptual framework. A simple example has been given of the way in which the recognition that matter exists in three forms, leading in our texts inevitably to an atomic conception, seemed to lead in Newton's day just as naturally to Aristotle's four elements – and after taking full account of the atomistic alternative. Many other examples of the complex relation between 'observation' and 'conclusions' could have been chosen: Copernican heliostatism; Newton's and post-Newtonian conception of forces; phlogiston; Darwinian evolution; ether theories; or the 'ultra-violet catastrophe'.

3   As so often in considering the problems of ethnic minorities, we find them rooted in more fundamental ones which affect everyone in society, even if their consequences may seem particularly disadvantageous for those minorities. Restructuring our school-science textbooks to make use of all that we have learnt about science through history and philosophy of science offers the prospect of helping to dispel not only the deeply negative image of non-Western cultures that is an almost incidental by-product of current ways of teaching, but of healing the divisive effects of the science–arts divide within our culture.

The 'discovery' approach has brought a novel excitement to school-science teaching. It does not have to be summarily discarded in the light of comments made in this paper. Rather, ways should be found of supplementing it in order to give pupils a richer sense of the background assumptions which guide the process of drawing inferences from 'observational' data. The history and philosophy of science may have an important role to play here. By enlarging our awareness of the very different ways of conceiving the world which have held sway in the past, and may do so again in future, it can help to mitigate the culturally-impoverishing consequences which flow from an unconscious arrogance towards the past and towards other cultures.

NOTES

1   Lewis (1977).
2   Newton's student notes are contained in Add. 3996 at the Cambridge University Library. The contents of the notebooks were first noticed by Hall (1984), and discussed more fully by Westfall (1962). The present paper draws freely on the sources used by Newton, and particularly on Magirus (1642) and Stahl (1652).

Thought in Action
Making Sense of Uncertainty in the Laboratory

David Gooding

Introduction

On 3 September 1821 a young scientist working in the basement of a house off Piccadilly made a current-carrying wire 'rotate' continuously about a magnet. This was the first time that an electric current had been made to power continuous motion. The apparatus emerged gradually from vague possibilities and the manipulation of magnets, needles, and wires (see figures 1 and 2).[1] The transition from possibilities to a real working prototype took at least a day. It is curious that, outside of history of science or the recollections of scientists themselves, little is ever said about truly exploratory experiments that enable scientists to elicit new phenomena. This is partly because it is difficult to recover the open-endedness and uncertainty of preliminary explorations made without a map, especially after you've mapped out the route from the vantage point of the final destination.[2] The discovery process is difficult to write about in ways that are acceptable to the image of science as a systematic and logically rigorous process.[3] The discovery I shall look at became one of the basic inventions of electromagnetic technology. As its theory became increasingly mathematical during the nineteenth century, so precision measurement in aid of the deductive testing of theory displaced the qualitative, exploratory work that had opened up the new fields of electromagnetic science and technology.

The view of experiment as the authoritative voice of nature was of course well established before the early nineteenth century. The inventor, Michael Faraday, was partly responsible for the idealized view of experiment as means of demonstration that found its way into science education during the latter half of the century.[4] By that time the name

Faraday was synonymous with experiment as a powerful means of discovering and demonstrating facts about nature.[5] It was the public and demonstrative side of experiment that found its way into science education, not the private, inventive, and constructive side. It is fair to ask whether invention and discovery have ever received their due in science education.

The image of experiment as an authoritative empirical test of theories still dominates our thinking.[6] Here I want to bring out the side of experiment that Faraday never disclosed. This will show how it became possible for Faraday to draw his first sketch of how the magnetic rotation might be achieved. This sketch appears at the bottom of figure 2. We can do this because he left us some remarkably complete records of the exploratory work that led to the first 'electro-magnetic motions', and because the experiments can be reproduced. First I shall consider the relevance to science teaching of this private and neglected side of real science. I shall not offer yet another case to be included in an already crowded syllabus, nor rehearse the arguments surrounding the Nuffield/PSSC approaches to experimentation. Instead, I want to encourage an approach to the teaching of some existing basic science and technology, in which pupils experience problem-solving and communication in a form that is not so different from what, according to many recent historical and sociological studies, happens in real science. This puts the uncertainty that scientists live with at the frontiers into the school laboratory. The purpose is to draw attention to the ways in which pupils' own agency and their interactions with others (including the teacher) affect both the way that they reach conclusions about how to interpret new experiences and the confidence they have in those conclusions. In this respect the approach differs from teaching schemes which make it a virtue to ignore both the individual, cognitive component of invention and the social, interactive basis of consensus.

Uncertainty in frontier science

I want to make three main points about the bit of frontier science recorded in figures 1 and 2.[7] The first is that Faraday made sense of electromagnetic phenomena partly through what he *did* with real objects in a concrete experimental situation. Inventive, playful thinking grappled with the natural world through an observer's hands. This led him to produce a completely new set of electromagnetic phenomena. Secondly, many of the actions he made in the course of these experiments were

Electromagnetic expts. with Hare's Calorimotor. To be re- ELECTRO-
membered that this is a single series? MAGNETISM.

1. Position of the expt. wire A*.
2. Positions at first ascertained were as follows

3. On examining these more minutely found that each pole had
4 positions, 2 of attraction and 2 of repulsion, thus

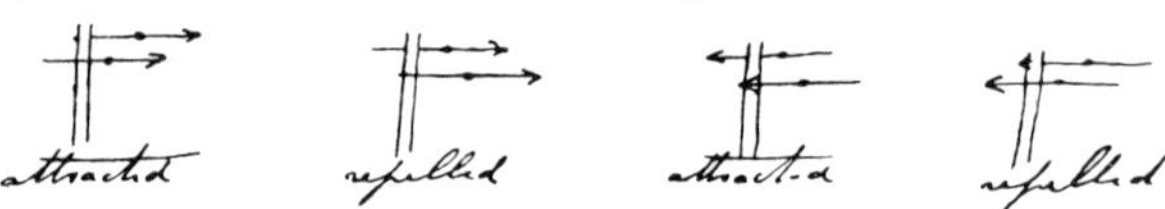

4. Or looking from above down on to sections of the wire

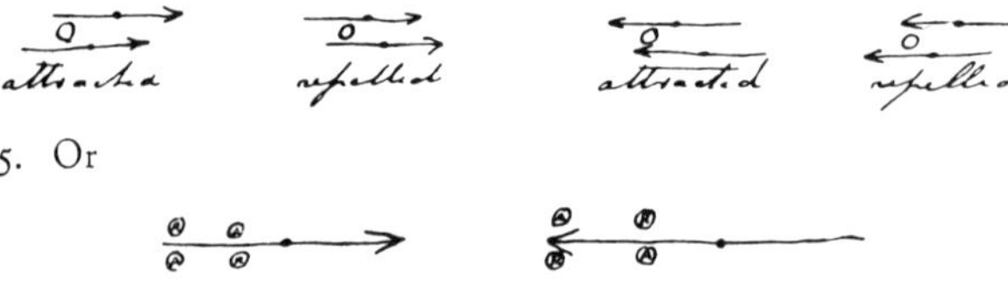

5. Or

6. These indicate motions in circles round each pole, thus

Hence the wire moves in opposite circles round each pole
and 'or the poles move in opposite circles round the wire. To
establish the motion of the wire a connecting piece was placed
upright in a cork on water; its lower end dipped into a little
basin of mercury in the water and its upper entered into a little
7. inverted silver cup containing a globule of mercury; the arrangement of battery poles always as at first. Magnets of different

Figure 1 Faraday's record of one of his earliest exploratory experiments on
electromagnetism of 3 September 1821, transcribed from his manuscript
diary by T. Martin, published as *Faraday's Diary* (see Martin (1932–6), vol.
1, p. 49).

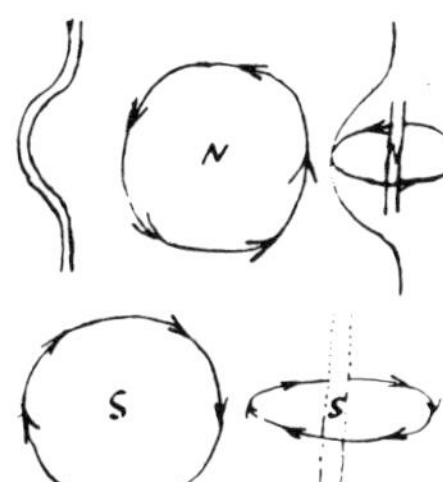

power brought perpendicularly to this wire did not make it re-
volve as Dr. Wollaston expected, but thrust it from side to side.
8, 9. The wire then bent into a crank form, thus, and by repeated
applications of the poles of the magnets the following motions
were ascertained, looking from above down on the circle described by the bent part of the wire, different Magnetic poles
shewn by letters, North pole in centre. The rod in the circle is
merely put there to shew the front and back part.

10. Magnetic poles on the outside of the circle the wire described*.

11. The effort of the wire is always to pass off at a right angle
from the pole, indeed to go in a circle round it; so when either
pole was brought up to the wire perpendicular to it and to the
radius of the circle it described, there was neither attraction nor
repulsion, but the moment the pole varied in the slightest manner
either in or out the wire moved one way or the other.

12. The poles of the magnet act on the bent wire in all positions
and not in the direction *only* of any axis of the magnet, so that
the current can hardly be cylindrical or arranged round the axis
of a cylinder?

13. From the motion above a single magnet pole in the centre
of one of the circles should make the wire continually turn round.
Arranged a magnet needle in a glass tube with mercury about it
and by a cork, water, etc. supported a connecting wire so that
the upper end should go into the silver cup and its mercury and
the lower move in the channel of mercury round the pole of the
needle. The battery arranged with the wire as before. In this way
got the revolution of the wire round the pole of the magnet. The
direction was as follows, looking from above down [see diagram].

Very Satisfactory, but make more sensible apparatus.

TUESDAY, SEPT. 4.

14, 15†. Apparatus for revolution of wire and magnet. A deep
basin with bit of wax at bottom and then filled with mercury,
a Magnet stuck upright in wax so that pole just above the surface
of mercury, then piece of wire floated by cork, at lower end

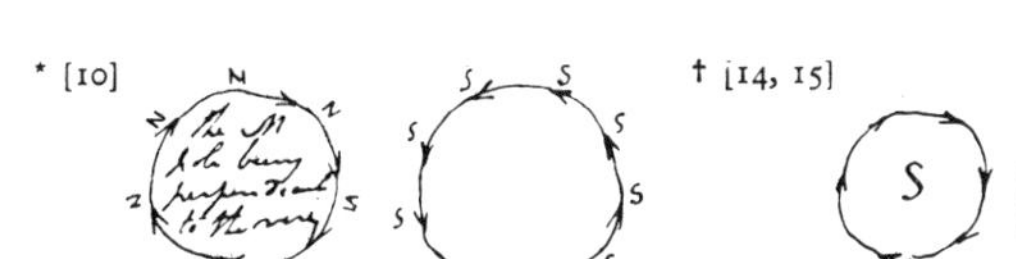

Figure 2   Continuation of the entry shown in figure 1. Faraday's first sketch
of an apparatus to produce continuous motion of a wire about a magnet
appears at the bottom left hand corner of the page.

not premeditated. Surrounded as we are by devices powered by electric motors, it is difficult to imagine what it must have been like to pursue what were then mere possibilities. Yet, until very nearly the end of the day he was articulating possibilities rather than testing predictions. Thirdly, like every other scientist before or since, Faraday needed to communicate the nascent, emerging 'sense' of what he was doing. He did this by making objects and drawings as well as by writing down results. We often fail to realize that, in the beginning, sense-making activity is essential to the very existence of a new phenomenon.

In this respect, new phenomena are human creations. They become scientific facts only after they have been reworked and rendered in a demonstrable, public form. Scientists at the frontier learn to find their way in uncharted terrain but must also invent ways of sending back despatches that will make sense to others. This is just as true for technology.[8] Innovators always have to communicate with others who have no first hand experience of just how unfamiliar and uncertain things are at the frontier. This is very clear in some of the entries Faraday made in his *Diary* for 3 September 1821.

My three points about agency, uncertainty, and communication are directly related. I say that experience is *made* because human agency is essential both to producing phenomena and to making sense of them. The scientist who fails to communicate what he finds fails to discover anything. The need for this sense-making is most apparent when the facts do not speak for themselves, when what they say is ambiguous or when the only available interpretation is incompatible with the accepted theories. One or more of these conditions *normally* obtains in research at the frontier of what can be thought, said or done.[9]

I have emphasized two kinds of inter-dependence: first, of thought and action, and second, of observers. The stage-setting needed to overcome the privacy of our own sensory experience and the unfamiliarity of new experience is inherently practical *and* social: making your own experience intelligible cannot be done in isolation from other would-be observers. To say that making-sense of a new phenomenon is an inherently social activity may sound tautologous – all the more surprising then that the discovery of new phenomena is so often presented in the same way as the discovery of new concepts: as having been the solitary activity of some lone genius. Art historians are not this naive: even the most individual and reclusive of artists is acknowledged to have completed his or her work only when an audience has been found (or created).

Uncertainty in the school laboratory

The problems of interpretation and communication face anyone who is trying to show something new in experience (whether in science, technology, or the arts). Then it exists for pupils as much in the science laboratory as it does elsewhere. Whenever they are asked to describe and interpret some phenomenal process new to them, school pupils are 'lay-observers', just as Faraday was when he began the work sketched in figure 1.[10] Pupils typically resolve uncertainties in two ways: by referring to peers (to each other or to a neighbouring group in the laboratory) or to authorities (texts or teachers). Scientists are often confronted with uncertainties of a sort that would be familiar to school science pupils. Positive use can be made of scientists' uncertainties to show pupils how scientists make sense of new aspects of nature and how they reach consensus about what their experiments produce. I am not suggesting 'guided discovery' in which pupils become 'real' scientists for an hour or an afternoon. I think that my example will have more to say about *the way* that particular material is presented. At the very least, pupils should be aware that scientists also find themselves in uncharted terrain, that they must learn to define and refine problems in order to solve them, and that this learning can itself take time.[11] It is just as important to understand how a scientist like Faraday reaches the point where he can draw the diagram of the apparatus in figure 2 as it is to see how he gets from that sketch to the demonstration device as published in figure 3. We can do this by looking closely at Faraday's record of his changing procedures and apparatus.

Faraday's Electromagnetic Explorations of 3 September 1821

Faraday's investigation of electromagnetic rotations was part of a flurry of activity engendered by Oersted's discovery of electromagnetism. Humphry Davy had taken up the phenomenon as soon as it became known in the autumn of 1820 and Faraday had assisted with many of his experiments. At this time he had yet to establish his reputation as an independent research scientist. During 1821 Faraday began to include his own investigations of electromagnetism in the notebook in which he recorded most of the experiments he was required to do in

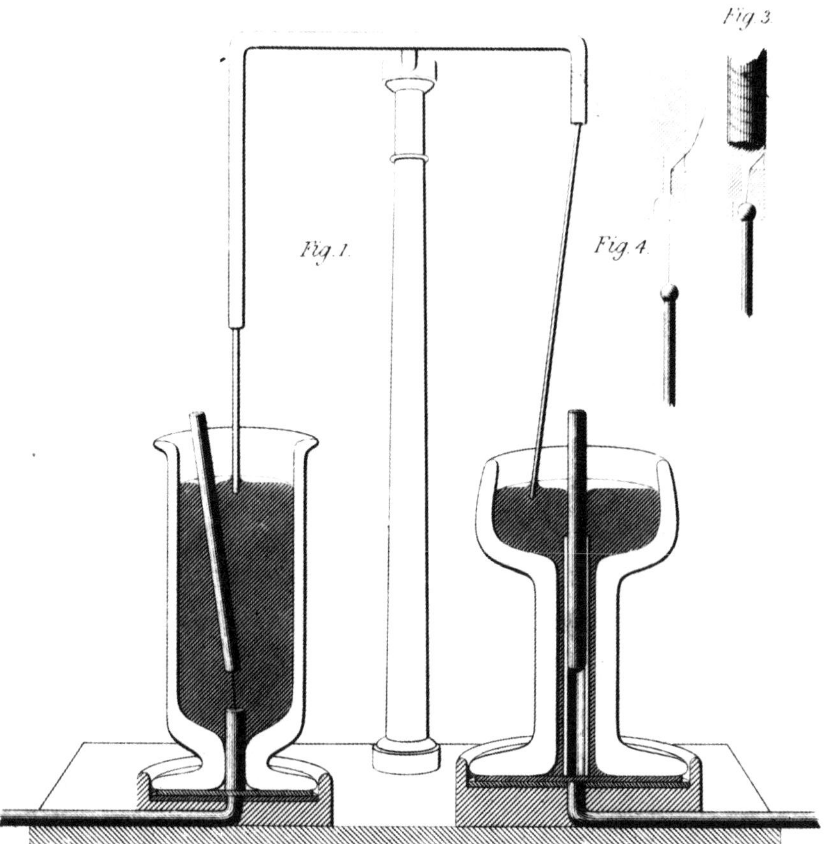

Figure 3 This engraving shows a version of the rotation apparatus built later in 1821 to demonstrate (on the right) the original discovery of magnetic motion and (on the left) the 'rotation' of a magnet around a vertical wire (from Faraday's *Experimental Researches in Electricity*, vol. 2, plate 4).

the laboratory of the Royal Institution. Figures 1 and 2 show Thomas Martin's transcription of his notes of an investigation of the behaviour of magnetized needles near current carrying wires, made on 3 September 1821. This was one of his earliest independent investigations of electromagnetism.

According to Faraday's record, the experimental wire was first positioned vertically so that it could be moved more freely around a needle. In the early experiments the needle was used both vertically and horizontally, suspended on a thread by one end or at its middle. Because of the excitement and the confusion generated by Oersted's account, Faraday was particularly interested in tendencies of the needle to approach or recede from the wire. He represented these visually, as properties of position. He then made a 'more minute' observation of the behaviour of the needle from above and from the side. These observations seemed to indicate that each pole of the needle had not one, but two positions of attraction and repulsion (see paragraph 3). These results are re-stated pictorially in his paragraphs 4 and 5. Finally, at paragraph 6, he gave a pictorial summary which appears to compress all of the preceding results into a single pair of diagrams.

It is unlikely that Faraday was looking for a well-defined effect. Like Lewis Carroll's Red Queen he had as many as six possibilities in play. (Unlike her, he had few preconceptions about which was impossible and he continued long after breakfast.) None was clearly formulated. One of the most important points to emerge from the close study of this sort of experimentation is the re-formulation and refinement of possibilities through the play of material objects and their properties. Since his expectations were not clear enough to lead him more than a little of the way it is not surprising that Faraday did not get a well-defined effect. Observation in science is rarely as straightforward as it appears to be in the definitive published accounts which are written after innovators have learned how to see. Recent attempts to reproduce some of his observations gave several insights into the procedural aspect of what Faraday says that he did. They also brought out the sheer difficulty of seeing what Faraday records that he saw, even after considerable practice.[12]

When we tried to repeat these first experiments we found it is extremely difficult to get a needle to remain still near a wire, especially when it is carrying a strong current. (The needle twists, due to tension in the thread by which it is suspended; it bobs about in convection currents; it interacts with earth's field, and so on.) The biggest difficulty is that it tends to touch the wire and this immediately alters its magnetization. Such practical difficulties are solved by changes in procedure; thus we found it easier to allow the needle to settle down, then move the wire around. Even so, we never did observe nicely ordered circles such as Faraday drew in paragraph 6 of figure 1. We also found that needles do not behave as if their magnetic poles are at their extremities (where most textbooks place them). The needles appear to be attracted or repelled at points a little way in from their tips. Faraday could easily have verified this behaviour using iron filings to map the field of the needle. If he did so, he did not think it worth recording. Although he had little theory to guide his experiments, he had no textbooks to mislead him either.

Creative uncertainty in the laboratory

I want to draw two main points from these examples. The first is that even having prior knowledge of a 'correct result' does not necessarily *help* you observe what you expect. This underlines the point that a great deal of skillful interaction with the phenomenal world is needed as well. Skillful manipulation of real objects and mental images of them was essential to the very observability of what Faraday recorded. Yet *it was not a part of what he recorded*. He recorded and later published only the stable residue, that is, 'rotation' as a natural phenomenon. Do we teach our children to disregard their own constructive activity by training them to record only 'objective phenomena' or what seems to take place 'out there'? I suspect that sense-making activity is written out of pupils' accounts in the same way that it is written out of scientists' accounts, as they move from exploratory, inventive work towards demonstration and proof. Surely there is an opportunity to draw pupils' attention to the ways in which they construct meaningful interpretations? Of course the opportunity is difficult to realize, given the need to use laboratory time to cover a syllabus and to teach practical skills. These pressures mean that pupils' uncertainty about their own procedures and results is often resolved in the way that Joan Solomon describes: 'the general atmosphere of unsureness [in a school laboratory] is not resolved by the experimental results, but only by the teacher's final summing up.'[13] Here is where the laboratory experience coincides with the finding of many historians and sociologists of science that experimental results are not, of themselves, compelling or decisive: replace the 'teacher's final summing up' by 'the judgement of the scientific community' and you have the analogous situation for experimental scientists. (The pressures on teachers often have their counterpart in real science as well.)

This brings me to my second point, about uncertainty. Not being so sure of a correct answer is not just an occupational hazard of life at the frontier: uncertainty is *necessary* to innovation. It should be put to positive use in the laboratory or classroom. Although preconceptions do not necessarily *prevent* you observing what you do not expect (as shown, for example, by the 'discovery' that the poles are not at the tips of the needle), the ability to be receptive to experience – to live with phenomenal ambiguity – is essential as a space for creative play with possibilities. Elspeth Crawford has drawn attention to the value to Faraday of living with ambiguity.[14]

For many teachers the line between ambiguity and plain chaos will seem fine, perhaps too fine to be drawn (let alone maintained) in the classroom. If uncertainty is resolved when the teacher intervenes too soon, then an opportunity to show how interactive processes of negotiation can lead to consensus about observations is lost. If the teacher does not intervene, then the demonstrative value of an experiment can be lost. These concerns should not disguise the fact that expertise and authority function to underwrite (or undermine) consensus in real science as well as school science. If pupils never come to appreciate the ambiguity and uncertainty of new phenomena, how can they understand that scientists also face this problem, let alone how scientists turn uncertainty into objective natural fact? If it cannot be enjoyed, uncertainty is still worth enduring, so that pupils' attention can be drawn to their own need to resolve uncertainty about experimental results, in order to focus on the ways such uncertainty is resolved.

I would guess that practicability is not yet at issue here because current teaching practice draws pupils' attention *away from* both the constructive (or cognitive) processes and the social ones (of appealing to peers or to authority). After all, aren't the natural phenomena and the basic science the only important things? Such teaching offers students a false view of science which makes scientists out to be super-competent because they are able to extract authoritative answers from nature. The contrast between pupils own uncertainty and inexperience and the apparent ability and certainty of real science increases their distance from science. This contrast is so exaggerated as to be seriously misleading. Being realistic about sense-making and consensus-formation in science can improve pupils' perceptions of themselves as inquirers, in relation to real scientists. The inspirational value of heroic images of 'great scientists' should not be bought at the expense of a false view of science.[15] I hope that recognition of the importance of uncertainty in real-science observation and experiment will encourage creative use of it in laboratories. To be enjoyed (or endured), uncertainty needs to be recovered from the notebooks and correspondence of experimenters. Here history of science has something to offer to school science.

Thought in Action

Finally, in order to illustrate how Faraday's 'doing' was essential to imaginative thought about phenomena, I shall work through what it was physically possible for him to have done with wire and magnet. In paragraph 6 a verbal description is accompanied by a diagram of the apparent motions. 'These [effects] indicate motions in circles round each pole . . . ' He then makes what looks like a prediction: 'Hence the wire moves in opposite circles around each pole and/or the poles move in opposite circles round the wire'. This articulates an imagined consequence of Faraday's interpretation of the behaviours of the needle obtained so far. It appears to say that the *wire* should move in circles about the pole of a needle. But there's a problem. He could not 'indicate motions in circles round each pole' with the arrangement drawn here because a needle does not approach or recede from a wire if, as shown in the diagram, the two are in the same plane.

Should the diagram be read differently from those above it? Here is one of those occasions when we need to avoid the influence of hindsight. To anyone who knows the 'right answer' it will seem self-evident that the circles in the diagram at paragraph 6 should be 'read' as a three-dimensional picture in which the circles are perpendicular to the wire and represent the path described by the tip of a wire moving about the pole of a horizontal magnet. But this interprets it in terms of what *we* now know, thanks to what *Faraday* had not yet found out. The diagram of paragraph 6 is just as readily interpreted as a 'summary' of the preceding results. It depicts successive positions of needles obtained in a lengthy sequence of operations. This compresses many actions into one temporal instant, moving from actual operations to a visual image, that is, from the *temporal* sequence to *spatial* structure. The image of a circle can interpret the arrangement of needle positions (of paragraphs 2 – 5) as a continuous process only if one makes a further mental manipulation. This is that the needle (or magnet) and wire must be in the same, vertical plane. This is probably how Faraday came to imagine that the *wire* might describe circles instead of the magnetized needle.

Making observations

The image of the circle (figure 2) was now an important heuristic for subsequent experimentation. This change was motivated by practical as well as intellectual considerations: it was made to solve a problem of rendering what Faraday now expected to see. He had to describe in two dimensions an interpretation of something that was supposed to happen in three. Here we have a close interaction of exploration and interpretation; of physical manipulations with mental ones. This interaction extends right into the imaging and recording of experience.

Faraday was now close to realizing an effect resembling a continuous circulation of one thing about another. In what followed the manipulation of objects was essential to his realizing the desired effect. This involved teasing out continuous circular motion from effects which were at first decidedly linear. In order to 'establish the motion of the wire' Faraday placed a 'connecting piece [of wire] . . . upright in a cork on water' so that 'its lower end dipped into a little basin of mercury in the water and its upper entered into a little inverted cup containing a globule of mercury'. He now had a wire that could move about its axis. Magnets 'of different powers' were 'brought perpendicularly to this wire'. Although he had not, apparently, been expecting lateral motion, the cups containing the mercury allowed for enough lateral motion to suggest another effect worth investigating. Further exploration resolved this into another geometrical relationship. This did not yet involve circles, but rather a tendency of the wire to move in either direction along a line perpendicular to the axis of an approaching magnet. He bent the wire into a crank form 'and by repeated applications of the poles of the magnets, the following motions were ascertained, looking from above down on the circle described by the bent part of the wire . . . '

Faraday wrote these words to help himself interpret his drawings, so as to recover his procedures later. The need to include instructions on reading his own sketches shows how ephemeral and plastic this new experience still was. Making an observation involves a dialectic of acting and rendering. Faraday's verbal and pictorial renderings of his experimental trials are more like the sketches with which artists select aspects of experience and work these into a stable, coherent, and compelling rendering of a face or a scene, than they are like science-textbook notions of scientific observations. The standard distinctions we make between doing, observing, and recording divorce the process of conceptualization and imaging from that of acting and making sense of actions and their outcomes.[16]

Articulating practice

The crank swung around so as to strike the magnet, but by moving the magnet quickly out of the way and reinserting it he could get the crank to keep moving. These operations would produce the motion recorded in the *Diary* as a ring in figure 2. If the ring represents the circle described by a point on the rotating crank, rotation would be produced by 'repeated applications of the poles' at points indicated by the letters 'N' and 'S' around the rings in the figure. By the time that he wrote in paragraph 11 that 'the effort of the wire is always to pass off at a right angle from the poles', he had explored the phenomenon sufficiently to realize that, 'when either pole was brought up to the wire perpendicular to it and to the radius of the circle it described, there was neither attraction nor repulsion'.

In paragraph 11 Faraday interprets his own exploratory *behaviour*. He interprets not only what he sees 'out there' but also what he has been doing. This interpretation articulates something realized implicitly in the crank variation of the moving wire experiment. Faraday now knew that a wire crank would try to rotate, as it were, around the pole of the magnet, if the pole were sufficiently close to the axis of the crank (not outside the circles). Such motion was a physical impossibility: if the magnet is in this position the crank must collide with it. As we saw, to obtain continuous motion he had to move the magnet out of the way and reinsert it as the crank went by.

He had not yet produced continuous motion of a wire about a fixed magnet unaided by active human intervention, as his diagram implied should be possible. He wanted to produce a *natural* phenomenon, one that did not involve human intervention. There was another solution. The little motions of paragraph 7 showed that an effect existed with the magnet 'perpendicular to' the wire. The crank variation showed that the pole needed to be close to the axis of motion of the crank or wire. Faraday had also learned that the magnet need not be in the same plane as that of the circle described by a point on the crank. So, both the magnet and the wire could be vertical, as they'd have to be in order to bring the pole inside the 'circle' of continuous motion. He inferred 'from the motion [of the crank] above' that 'a single magnet pole in the centre of one of the circles should make the wire continually turn around'. By the time he wrote down paragraph 13, Faraday had realized that there is another way of getting the wire to move: he could straighten the wire but keep the magnetic pole 'inside' the axis of possible motion of the wire by turning the magnet through 90°, so that magnet and wire have a common axis. The circles of paragraph 6 would now represent the actual path of the *end* of the wire if this formed part of an electric circuit.

At last the problem was well defined in the way that scientific problems are defined for publication. Possible solutions were suggested by what he now knew, both conceptually and practically. These possibilities were also more highly constrained by this knowledge, as his sketch (figure 3) shows. Yet rotations, like the 'positions' of the needles with which we began, cannot simply be observed like mountain ranges or the moon.[17] Faraday was well aware that, to make the phenomenon real, he had to make it public and, therefore, accessible. His concluding entry for the day read: 'Very satisfactory, but make a more sensible apparatus'. One version of this demonstration device is shown in figure 3.[18] Too often the story of the demonstration experiment is the only one told in textbooks. One service history of science can offer to school science is to make the private lives of such experiments more accessible.

Conclusion

I have made two main claims about 'frontier science'. The first is that the results scientists present as bold, self-evident facts about nature have been paintstakingly elicited, shaped, and made visible through the invention of methods of observation and representation. Then they have been disseminated through the invention and perfection of experimental techniques. These activities are as much about creating and communicating meaning (or sense) as they are about disclosure and proof. The second point is that learning to elicit phenomena in such experiments involves many manipulations of objects in real and imaginary space. Faraday did not derive procedures or their outcomes from the propositions of some theory. Nor did he stumble upon the prediction of paragraphs 6 and 13 by trial and error. This sequence of operations is a fine example of practical intelligence.[19] My interpretation of

Faraday's record illustrates that observation and experiment are *active*.[20] The close interaction of thought, action and belief also shows that the Cartesian distinction between 'thought' and 'action' works against recovering the making of experience just as much as hindsight ('knowing the answer') does. These aspects of discovery are important to scientific innovation. We need to consider how science teaching can draw attention to them. Certainly teachers should not dismiss them as irrelevant to the production of objective knowledge. This calls into question the theory of knowledge on which most views of science are based.

NOTES

1　These are reproduced from Martin (1932–6), vol. 1.
2　John Tyndall remarked of Faraday's researches it has been 'a heavy labour to master them' because they 'are so voluminous, their descriptions are so detailed, and their wealth of illustration is so great . . . ' (Tyndall (1884), on p. vii of the preface to the second edition, which was published in 1869).
3　On the misrepresentations of discovery processes see Medawar (1964).
4　In addition to his reputation as a master experimentalist there was the image of Faraday the discoverer, created by Tyndall, Bence Jones (1862), and others, and of course the evidence Faraday himself gave to a public schools commission on science education in 1862.
5　Gooding (1985a).
6　Gooding et al. (1988).
7　These points are developed in Gooding (1989).
8　As a distinction based upon purposes and end use rather than the process of discovery and invention, the usual distinction between science and technology is not applicable to 'frontier' research. The ambiguous 'scientific' or 'technological' identity of the present invention is discussed in Gooding (1985b).
9　Note that I am concerned here with resolving and communicating uncertain interpretations of phenomenon, not with acceptance by an audience already committed to some theoretical interpretation. (The construction of consensus about theoretical concepts is discussed by Pumfrey in this volume.)
10　There is no point in trying to win others' assent unless you can be sure that they can see just that aspect of phenomenon you intend. The distinction between cognitive and rhetorical strategies in experiment is developed in Gooding (1986, 1988, 1989).
10　For 'lay observers' see Gooding (1985a).
11　On the time that discoveries can take, see Gruber (1981).
12　This was carried out with Miss Jane Leigh and Miss Susan Perrett as part of their final year Physics project at Bristol University. The work is reported in greater detail in my 'Techniques from texts?: recovering the micro-structure of real experiments' (unpublished).
13　Solomon (1980), p. 112.
14　Crawford (1985).

15 Heroic images are easily misused (as I believe Mrs. Thatcher's image of a self-educated Faraday was, in the transcript of a BBC-TV interview with Russell Harty for 'Favourite Things', broadcast 26 July 1987).

16 Gooding (1988).

17 Our repetitions showed that getting the wire to describe circles about the magnet can be difficult even after you have the right configuration of magnet and wire.

18 See Faraday (1821).

19 In *The Concept of Mind* Ryle pointed out that practice can be intelligent without being pre-meditated; the 'intelligence' consists in learning through doing, where each repetition modifies the next. Ryle (1949), chap. 2. See also Harrison (1978).

20 Harré (1981) distinguishes between active experiment, passive observation and exploratory investigation. He identifies active manipulation with 'true experiment' (pp. 21–3). I have emphasized the active, exploratory character of Faraday's investigation because I think that this is essential to all experimentation and much observation. See Gooding (1989), chap. 2.

The Concept of Oxygen
Using History of Science in
Science Teaching

Stephen Pumfrey

Problems of Discovery

In this paper I open up two well known and interrelated problems surrounding the 'discovery' of oxygen. The first is historical. Since Thomas Kuhn's *Structure of Scientific Revolutions* research has made the notion of a definitive moment of discovery for oxygen untenable.[1] Indeed many historians of science have become sceptical of all discovery stories.

The second problem is science-educational, and concerns the 'guided discovery' of oxygen by young secondary school students.[2] During my short period of teaching mixed ability, integrated science I was able to consider the problems of guided discovery from the perspective of a historian of science. The general weaknesses of guided discovery have been well explored,[3] and many teachers have voiced their own dislike of the false picture of science it conveys. In this discussion of oxygen, I wish partly to add to the criticisms, but also to save a place for guided experiments by interrelating students' problems to the historical problems of its development. Guided discovery needs to be saved in some form, certainly not because it is an accurate recapitulation of the scientific process of discovery, but because it allows such good student participation in concept formation.

To return to the history of oxygen. What would we need to locate its discovery and discoverer? You will be familiar with some of these elementary points. We don't want to say that the discoverer was the first person to hold a sample – that would be the first bottler of air. We might insist that discoverer had awareness. This rules out the Swedish apothecary Carl Scheele, who probably prepared the first purified sample, but thought it was common air. Joseph Priestley realized he had isolated a remarkable gas when he obtained it from mercuric oxide in November 1774, though at first he thought it was nitrous air. Six months later he knew it was purer than 'common air', and his qualitative tests of this were those we use today. Yet, being a phlogiston chemist of course, he died insisting that there was no oxygen: it was 'dephlogisticated air'. We are left with Antoine Lavoisier, who prepared oxygen from Priestley's account, and thought it was common air until Priestley indicated its special properties, from which he developed the oxygène theory.[4] But not only was he not the first to isolate it, he does not meet our criterion of knowledge. Lavoisier's oxygène was not our oxygen. He propounded it as the principle of all acids, which Priestley knew it was not, and he held the gas to be a *compound*, of the oxygène principle and the heat fluid caloric.[5] As Lavoisier's very different and counterfactual concept of oxygen became dominant through the diffusion of French chemical practice, textbooks, and nomenclature, it was developed towards our own understanding, while phlogiston disappeared from nature.

Lavoisier's and Priestley's supporters possessed, in some form, all the evidence we lead our pupils towards today. It is clear that they disagreed over the interpretation of indecisive evidence. Historians conclude from these circumstances of the 1770s, first, that oxygen was not a natural fact to be discovered, but a theoretical entity to be interpreted and constructed (you may well agree). We also argue that Lavoisier's success cannot be attributed to the evidence alone (which was disputed), but to the social circumstances of eighteenth century debates in chemistry. I will mention some of these later.

Before we go into detail, I want to emphasize that the experimental support for oxygen was not self-evident in the 1780s, and nor is it today, *unless* we already view the world from the perspective (or in Wittgenstein's more apposite term, 'form of life'[6]) of oxygen theory. Lavoisier and we do; Priestley and our pupils don't.

Images of Science: Student, Teacher, Historian

Moving on to the classroom situation, I want to explore how lessons drawn from the problematic reception of oxygen in the 1770s and 80s can ease its problematic reception by pupils today.

The picture of the unsuccessful guided discovery class is painfully well known. Pupils heat some elemental copper metal and observe the formation of a black deposit of oxide. We ask them to make hypotheses about its formation. Embarrassed silence. If we are fortunate, someone will suggest one of the standard and easily disproved explanations, such as 'it comes from the flame' or 'it's like soot'. More often we have to suggest them or cope with difficult hypotheses such as 'the heat has burned and changed the copper'. After these have been discounted, by heating in a boiling tube and so forth, there is another demeaning silence until someone notices a vacuum pump in the corner. Since pupils know that they are engaged in 'cold' or mock science, not 'hot' discovery science,[7] they reason that the teacher knows the right answer, that the vacuum pump must be part of it, and so suggest that air might be important. (Does any pupil ever suggest a role for the air who is not already acquainted with calcination?) Using the pump, the teacher (*deus ex machina*) shows that air is essential, and then a series of dubious volumetric and gravimetric experiments confirm that a portion of the air, to be called 'oxygen', combines with metals to produce oxides. Of course, the whole guided sequence is ruined if a smart alec exclaims 'it's oxygen' during the very first discussion.

There are, I consider, three flaws in this 'guided discovery' approach, which an historical reformulation can correct. First, it is very damaging to pupils' self-esteem – their self-image as potential scientists. They are rarely able to develop hypotheses. When they do, it is rarely feasible to incorporate or test them, and if they are tested, they are usually wrong. My aim is to use history to give them a fighting chance.[8]

My second point is that the approach presents an alienating image of science itself. The ideology behind guided discovery teaching, or indeed the experimental scientific method, is that nature will yield clear answers to the good scientist. He poses the right questions and tests them in the right way. Historically science did not proceed like this, least of all in the case of oxygen, except in cleaned-up textbook distortions. Our students become alienated because they conclude that they do not possess the mentality to be successful at science. They cannot solve the puzzle.[9]

This brings me to my third, final point – that discovery is not possible, even if guided, for the evidence we give them is not compelling. I noted that experimental evidence did not resolve the historical debate, as we shall see.

Recent writers like Wiebe Bijker have noted that this traditional image of scientific knowledge as old (accomplished), alienated (done by geniuses), and almighty (the answer is always found) is partially responsible for the poor classroom image of science, with falling numbers and motivation, especially among girls.[10]

How can recent history help to close this perceptual gap about experiments which we agree are suggestive and persuasive, but which too often our pupils do not? The problem, I suggest, is that teachers are trained to give a *member's account* of these experiments, when what they need is a *stranger's account* from an historian. Let me explain these terms.[11]

Teachers generally, as part of the scientific community, have been inducted into the values and assumptions of science. Crudely put, 'science is objective, knowledge is cumulative, the experimental method works, oxygen is real'. A member's account of the past extracts from it what is useful to the present, such as good examples of experimental discovery. It must be said that increasing numbers of teachers and scientists are questioning this ideology. Historians however try to come to science as strangers. We pretend not to see the self-evidence of its assumptions. The members' current criteria may make oxygen natural and phlogiston fictional, but this will not help in *historical* explanations. Oxygen was not taken for granted in the 1780s. We examine that which is taken for granted to see why it is taken for granted. Since late eighteenth century chemists could not know the outcome of the debate, we look to social and personal factors to explain their actions.

One very useful way for the historian to build up his stranger's account is to look at past controversies, such as that over combustion. By treating charitably the opponents of victorious claims, like the phlogistonists, we are more able to know what was necessary for the victorious knowledge claim to become a natural fact.

The relevance to science education of strangers' accounts is, I hope, obvious. Our students come to a particular piece of science not as members but as strangers. They are unlikely to share all of the assumptions that are necessary to see the experiment as we see it. (If they do, then they should be doing something else.)

Good teachers have always paid attention to the conceptual changes they are trying to bring about in their pupils. But playing the stranger to all of our assumptions is difficult. Rosalind Driver's work helps to identify pupils' conceptual starting points,[12] but we also need to stand outside our end point. I am proposing that history of science enables us to do this. Using it, we can foresee many of our pupils' conceptual deficiencies.

The Interrelation of Historical and Classroom Problems

Joseph Priestley (1733–1804), the political radical and non-conformist minister, was also England's leading chemical philosopher. As a phlogiston chemist, Priestley took organic combustion as his paradigm, in which coal, for example, consists of earth plus phlogiston (which makes it flammable). On combustion the phlogiston is liberated, creating light and heat, and leaves a lighter, earthy residue of ash.[13] So the equation for the anomalous process of metal calcination is:

$$\text{metal} \;\rightarrow\; \text{calx} + \text{phlogiston} \uparrow$$
$$\text{(earthy + calx + phlogiston)} \qquad \text{(earth)}$$

Notice the obvious point that Priestley, and indeed the chemical tradition up to Lavoisier, considers metals to be compounds, and oxides to be elemental earths or calxes. The calx is a residue left by the escape of phlogiston, whose presence in all metals is evidenced by the common metallic properties of lustre and combustibility not shared by their calxes. (Rosalind Driver has found that children tend to begin with an organic model of combustion. They see burning as the liberation of something fiery, which leaves an ash. She notes the similarity of this to phlogiston theory.[14])

One of Lavoisier's fundamental conceptual transformations, of course, was to reclassify 'elemental' earths as compound oxides, and 'compound' metals as elements.[15] In the copper experiment, students *must* see their copper and deposit Lavoisier's way if they are to make successful hypotheses. We do see it, and teachers tend to talk about a black deposit forming on the copper, as though this were a neutral description – why not an ashy residue being left by the copper? Yet many pupils do not see the reactant and product our way, even if they have been previously taught about elements and compounds. Here is a fundamental conceptual block.

There are many more such assumptions of the member's account which we need to unpack for our students. Consider the role of the air. One did not need to be Lavoisier to know that air was essential both to respiration and combustion, but how? Prior to the eighteenth century, and in particular to Black's work on *magnesia alba*, philosophers ascribed no chemical, only physical, properties to air.[16] Pneumatic chemistry, gaseous chemical changes, was an achievement of the mid-eighteenth century, in which Priestley was pre-eminent, isolating some 13 chemically distinct 'airs'.[17]

Although Lavoisier argued that a portion of air combined with metals on combustion, Priestley knew perfectly well that air was essential to combustion. He knew also that the volume of air decreased by some 20%, and more if 'dephlogisticated air' was used. He explained this by giving airs the role of phlogiston receivers. On combustion, a closed volume of air became more and more phlogisticated until it could hold no more, and combustion ceased. Dephlogisticated air was obviously better. Volume decrease could be explained partly by the absorbtion of 'fixed air' (carbon dioxide). But phlogiston could also have the effect of decreasing the repulsion of air particles, resulting in a progressive reduction of volume.[18]

This review reveals more of the concepts our students must have internalized if they are to make successful hypotheses, or interpret the experiments correctly. Do they think of air as a mixture of substances with different chemical properties? Do they recall that some gases, such as carbon dioxide, react with solids? Do they interpret gaseous volume decrease definitively, in this case (but not others), as a decrease in mass? Do they think that the air gives or receives something in combustion? Many pupils I have talked to who hypothesized a role for the air, some of whom could even mouth the word 'oxygen', nevertheless believed like phlogistonists that combustion ceased because the air became 'poisoned' (which is, of course, the case in respiration). How would you show them otherwise?

Finally, we come to the gravimetric evidence discussed by the French and phlogiston chemists. The anomalous gain in weight of metals on combustion, when most substances appeared to lose weight, was not persuasively argued until the work of Lavoisier's colleague Guyton de Morveau in 1772.[19] The results were difficult to reproduce, and even more difficult were Lavoisier's claims to have detected weight similar increases for sulphur and phosphorus. Equally contentious was his claim to have shown that, in closed systems, the weight increase of metals matched the weight decrease of the air. Thus, as with school experiments, the results were not conclusive.[20] Where they were, as with the weight increase of calcined metals, phlogistonists were able to argue that phlogiston might have negative weight, or more subtly that this was due to absorbtion of water and fixed air, or even that weight changes did not matter.[21]

Priestley explored this view, for reasons which stem from his theology.

He was concerned to show the control of God (who was spirit) over material nature. Priestley argued in his *Disquisition Relating to Matter and Spirit* that, since we only know of matter through the forces or 'powers' it possesses, we can and should deny that there is any solid matter existing behind these powers of attraction and repulsion. Priestley conceived of matter as a set of various 'spiritual' powers. Once the mass or volume of a substance depends upon powers which (by analogy with chemical activity) can change a priori during a reaction, volumetric and gravimetric arguments cease to carry as much weight. I shall mention below why the French attitude was different.

As strangers we are not allowed to dismiss Priestley's view. In any case, our stranger-pupils share parts of it. We should ask whether they find the increase in mass of calcined metals significant, indeed whether they understand the principle of the conservation of mass.

Indeed, do they think that air or gases have mass at all? A century before Lavoisier, before Boyle's pneumatic experiments of the 1660s, the Aristotelian common sense notion that air has levity, not gravity, was dominant.[22] You may have been surprised, like me, with pupils' Aristotelian views of air. And they are astounded to find that air is heavy. Yet it is essential to their understanding of oxygen that pupils perform, or be reminded of, experiments to show that air has weight.

I think this brief historical perspective on our piece of classroom activity makes it clear why it is an unsuccessful and alienating experience for many pupils. In order to be guided towards the intended and supposedly unambiguous hypothesis from the experimental evidence available to them, they would have to bring to bear the following pre-existent concepts:

> Gases have mass.
> Gases have chemical properties.
> Gases combine with solids.
> Copper is an element.
> Its calx is a compound.
> Invisible particles are absorbed, not released, in this combustion.
> Mass is conserved.
> Mass considerations are more important than philosophical ones.

In short, in order to see the experiments successfully, they have to negotiate many of the disputes of eighteenth century chemistry. In particular, they have to see the experiments from the perspective of Lavoisier's oxygen theory, in order to arrive at Lavoisier's oxygen theory.

We do not always notice this catch-22 because we inhabit a Lavoisierian form of life. While this historical analysis exposes further the fraudulence of guided discovery it suggests, through our list of necessary tacit assumptions, a solution to the problem, which I pursue in the outline teaching scheme below.

Establishing the Concept of Oxygen: In Eighteenth century France

The above evidence, most of it well known to historians, simply confirms two standard axioms of recent science sociology. The first states that all experiments are open to dispute and interpretation. I have reminded you that the evidence we lead pupils to was well known, but not decisive in the eighteenth century. The second states that for one interpretation (among several) to seem natural one has to share the form of life of that interpretation: that is one has to acquire the skills, training, values, and interests of that school of interpretation.[23] This latter point conveniently allows me to digress onto how historians explain the contemporary success of Lavoisier's.

Opponents made fundamental criticisms – as Kuhn pointed out, all theories are born refuted. Oxygène was supposed to be the universal principle of acidity, yet several acids did not seem to contain it. Some yielded it through electrolysis in the nineteenth century, but Priestley's favourite counter-example of muriatic (or hydrochloric) acid never did.[24] It was also supposed to be the universal principle of combustion, yet other substances, which we guess were chlorine and carbon monoxide, also supported it. It was also supposed to replace the nonsense of the weightless phlogiston, yet it introduced the new element of caloric, the weightless fluid of heat.[25] Light, incidentally, also featured in the table of elements in Lavoisier's *Elements of Chemistry* of 1789.[26]

One more recondite problem – since the work of Cavendish in 1766 it was known that the gas liberated by metals (hydrogen) and organic substances (this was perhaps methane) when placed in acid was itself inflammable in air. The gas was therefore called 'inflammable air' or 'phlogisticated air', and its existence was counted as powerful evidence in favour of the phlogiston theory. Lavoisier seized upon Cavendish's synthesis of water from 'phlogisticated air' with 'dephlogisticated air', naming the former air 'l'hydrogène'. Unfortunately, the synthesis produced not pure water, but an acid solution, which phlogistonists could explain more easily.[27]

A more contextual explanation of the French success is tangential to this paper, but I'd better suggest answers to the questions I've raised. Here are the three most important factors which recent historians have adduced.

Firstly, *systematisation*. Lavoisier's *Elements of Chemistry* put order into the practice of chemistry, which had degenerated under the eighteenth century identification of new substances, especially airs. Lavoisier actually announced his intention to revolutionize chemistry before he began. I intimated that little of the order survives, but it gave chemists a textbook, theory, and laboratory procedure of analysis with which to advance their discipline.[28]

Secondly, different groups had *conflicting aims in chemistry*. French chemistry was analytical. From laboratory procedures Lavoisier bequeathed his list of 33 elements. His physical analysis, particularly the use of gravimetry, mirrored the approach of the much more prestigious group of French physicists, headed by Laplace and Coulomb. French chemistry, already well organized, had the support of the Academy of Sciences.[29] This contrasts not only with the disorganization of the phlogistonists, but with their aims. Priestley was horrified by Lavoisier's analytical approach. In the tradition of English experimental philosophy, he was looking for a small number of active principles, like phlogiston, which would demonstrate the economy and unity of God's creation. Priestley objected that the imposition of an areligious analytical theory was 'establishing a reign to resemble that of Robespierre'.[30]

Finally there is the dimension of *propaganda*. Just as Lavoisier planned his revolution, so his group planned its success. We see this in the multi-authored *Method of Chemical Nomenclature* of 1787[31] from which we get our very theory-laden terms like oxide, and sulphurous and sulphuric acid, where the suffixes indicated the relative amounts of oxygen in them.

The Scottish chemist John Keir protested that 'the language is that of the anti-phlogistic sect . . . It cannot be considered the general language of chemistry.' Lavoisier knew this. 'It will be necessary', he wrote, 'either to reject the nomenclature, or to follow inevitably the path which it indicates.'[32] We have inevitably followed that path.

Establishing the Concept of Oxygen:
In the Classroom

I am not, of course, suggesting that the above contextual issues can or should be presented to pupils. It is the aspects of the debate examined above which can be put to use. I would like to conclude by pulling together some suggestions, based on this discussion, for changing the way we teach oxygen. Recall that my analysis began from students' failure in guided discoveries, and a wide concern with their poor image of science and self-image as scientists. Thus the suggestions are informed by three aims:

1   to show scientific concepts are not natural and luminous even (especially!) to good scientists. Science is disputed and negotiated.
2   to let students feel that their ideas have value, by developing and legitimating them.
3   to give students a reasonable chance to make and test 'correct' hypotheses about combustion and oxidation.

The following points are exemplified in the sample scheme and materials listed in the Appendix. My collaborator on this scheme[33] and I do not present it as a good example (it is incomplete, trialled but once, and only partially successful), but it provides a starting point from which I invite comments and developments.

Our first concern should be, perhaps, to indicate that the concept of oxygen was not, and is not, manifest to all but the scientifically illiterate. It was disputed by good scientists. Thus the scheme is not introduced as 'the discovery of oxygen', but (rather melodramatically) as 'the battle over burning'. Introducing the disputants Priestley and Lavoisier carries the image of science as negotiated – with personal disagreement, argument over evidence, and a frequent absence of the obvious and natural. Several educationists believe this projects a more human image of science with which pupils can identify.[34] I was surprised how many pupils sympathized with Priestley's insistence on a religious dimension to the debate.

From this premiss, we might like to pick up on Driver's point that children carry a quasi-phlogiston model of combustion in their heads.[35] The scheme allowed them briefly to develop this, and explained Priestley's position. This has two advantages. First, they realize that their ideas are, and have been, scientific in a different way. Secondly, Priestley's position is an excellent shared perspective from which one can begin a conceptual transformation towards oxygen theory. Similar opportunities are available on the question of the gravity or levity of air. It should be noted that evidence from the trial of the materials and from other teachers' reports shows a danger that students find the phlogiston model too persuasive.

Thirdly, from a historical perspective we can isolate a full and necessary order of concepts which the pupils must have at their disposal. This point stands, I think, quite independently from the more contentious issue of the image of science. I refer you again to the list of concepts necessary to approach the copper experiment with promising hypotheses. The most important are that gases have gravity and chemical activity, and that metals are not compounds. Less important are an understanding of gravimetry, the conservation of mass and the 'anomalous' mass increase of oxidized metals.

It has been objected that these activities may be necessary in an ideal world, but would take up too much lesson time. There is some force in this, but many of the activities in the scheme are already part of the curriculum, but have simply been repositioned in order to make the concept of oxygen more accessible.

With all this effort put into one small part of the curriculum you'll not be surprised to hear that the crucial session on explaining copper oxidation went very well indeed. There were plenty of hypotheses, excellent tests and comprehension, and only a few puzzled phlogistonists at the end. Heuristically, then, using an historical perspective seems to improve students understanding of oxygen despite (I would claim because) the deliberate demystification of its luminous 'discoverable' quality.

I shall conclude with what I consider the most forceful objection to this kind of scheme. Science teaching and history of science are very different disciplines, with diverging aims in the accounts they give of scientific knowledge. This scheme is neither proper science, nor proper history – not proper science because of the time it devotes to discarded knowledge, and not proper history because the students are not given enough information on the social context of Priestley and Lavoisier. The historical elements are in some ways a 'rational reconstruction' of the debate, reconstructed for specific pedagogical aims. The scheme extracts from the past what is useful to the present, exactly the criticism I made of members' history of science. Given the constraints of the school laboratory, this is inevitable. Ultimately, the use of historical perspectives in school science can only be justified by results, in terms of student understanding of the nature of the scientific process. On this issue, the jury is still out.

APPENDIX A: SAMPLE SCHEME

Most of the following scheme was developed and informally trialled with mixed ability 12 to 13-year-old students of integrated science.[36] It is regrettably not possible here to reproduce the worksheets or analysis of the trial. References to 'NCS' are to experiments from Nuffield Combined Science (Nuffield Foundation, 1970). References to 'RCN' are to Revised Nuffield Chemistry (Nuffield Foundation, 1975).

THE BATTLE OVER BURNING

1 Historical introduction to ideas about fire and its importance. Priestley and Lavoisier's 'Battle over Burning'. Homework task: to find out about Priestley and Lavoisier, and why both died unhappily.
2 'Burning Questions'. Open-ended exploration into the effect of heating substances (NCS 2.2a). Informal groupwork to help students suggest their own hypotheses about which substances burn and why.
3 Cartoon introduction to simplified phlogiston theory. Compare with intuitive ideas. Show early section of ICI 'Oxygen' film.
4 'Does air really exist?' Establish that gases are real substances (NCS 4.1, 4.2).
5 'Does air help things burn?' General discussion. Burning candle in a confined volume of air (RNS A4.2a). Phlogiston explanations.
6 'Looking more closely at air.' Air as a mixture of substances. Making different 'airs'. Hydrogen, oxygen, carbon dioxide, and simple tests for them. Students invent their own names for the airs based on descriptive classification.
7 Priestley's religious interests. Degrees of 'goodness' of airs. Link to simple awareness of the 'oxygen cycle'. The success of the phlogiston theory.
8 Prediction of mass change on combustion. Test hypotheses (NCS 2.2c). Anomalous mass increase of metals.
9 Debate. Is the mass increase significant? Is it hard to explain? Lavoisier, the balance, the importance scientists give to mass changes. Can phlogiston theory explain these well? (Engineer a negative response.)
10 'Where does the extra mass come from?' Burning a candle on wooden float in bell jar over a water trough. Hypotheses for the volume decrease. (Priestley's defence is too complex for most students.) Testing the remaining gas. Reminder that gases have mass. Repeat with magnesium (difficult).
11 'Where does the black stuff come from?' Students to devise and test hypotheses for copper oxidation as fully as possible (NCS 2.2b). If necessary, introduce Lavoisier's basic theory (that some part of the air has combined with the metal).
12 Reinforcement of the ideas, suggesting the conservation of mass. Lavoisier's crucial experiment with mercury (second part of ICI film). Mass changes and

gas production on heating potassium permanganate (NCS 2.2g; NCS 4.3d). Burning candles (RNC A4.2b, A4.3). The percentage of 'active air' (RNC A3.1).

13   Burning other materials in air and oxygen (RNC A4.1). Mention of Priestley's use of mice in bell jars to test difference between 'common air' and 'oxygen' can raise questions of the experimental use of animals (and changing attitude to animal life since the eighteenth century).

14   The new concept of an element (in general, that there are many; specifically, the classification of metals and calxes). Students to review evidence for this. Lavoisier's nomenclature.

15   Final discussion. 'Did Priestley or Lavoisier "discover" oxygen'?

(It has been reported that role play simulations between Lavoisier and Priestley can work with able students. Less scientifically demanding role-plays portray the unusually prominent, though ancillary, part played by Lavoisier's wife in his work, and his trial and execution before an anti-scientific – and anti-capitalist – revolutionary court.)

NOTES

1   Kuhn (1962), pp. 69–71 and passim; for developments see Barnes (1982) and Lamb and Easton (1984), pp. 19–21. For a recent historiography of the eighteenth century see Crosland (1980). More generally, see Russell (1985).
2   See for example, Nuffield Foundation (1970).
3   For example Atkinson and Delamont (1977), pp. 87–108.
4   A concise account of the developments is Hankins (1985), especially pp. 100–9. The fullest, but most Whiggish, is of course Partington (1962), vol. 3, pp. 256ff; the debate is well presented through primary sources in the classic *Harvard Case Histories in Experimental Science* (Conant 1964), pp. 65–117. See esp. p. 77.
5   Lavoisier (1790/1965).
6   The concept is considerably developed in Bloor (1976).
7   The terms are from Atkinson and Delamont (1977).
8   See Solomon (1986), pp. 437–42 for a discussion of esteem and motivation.
9   Bijker (1983), p. 63.
10   Bijker (1983), p. 3.
11   They are better explained at length in Shapin and Schaffer (1985).
12   See, for example, Driver, Guesne, and Tiberghien (1985).
13   Conant (1964), pp. 70ff.
14   Driver, Guesne, and Tiberghien (1985), p. 158.
15   But by no means in the same sense as we use them. See Conant (1964), pp. 72–3.
16   Hankins (1985), pp. 81–93.
17   For some of these see Partington (1962), pp. 247–56.
18   See the illuminating articles by McEvoy (1978), pp. 1–55, 93–117, 153–75; (1979), pp. 16–38. On this point see McEvoy (1978), pp. 102, 110.
19   Hankins (1985), pp. 97–8.
20   See Partington (1962), pp. 385–6, and Conant (1964), pp. 84–6. It is clear

that Lavoisier fitted his results to his expectations.
21 For one version of the modified phlogiston theory see Conant (1964), p. 110; see also Priestley (1782).
22 Although people before Boyle had insisted on the absolute gravity of air, notably Galileo.
23 This approach is well described in Collins (1985), chaps. 1 and 2.
24 Partington (1962), p. 336.
25 Lavoisier (1790/1965), p. 1–25.
26 Lavoisier (1790/1965), p. 175.
27 At one point both Cavendish and Priestley believed that 'inflammable air' actually *was* phlogiston. See Partington (1962), pp. 312–14, 335–42.
28 Hankins (1985), pp. 106–10. As he notes, 'the rest of Lavoisier's Chemical Revolution had a more polemical character'.
29 A prime example of this is the co-operation between Laplace and Lavoisier to develop the ice calorimeter, which made caloric visible in the form of melted ice.
30 McEvoy (1978), 25, p. 34. The general difference in aims is made clear in section V, pp. 30–9.
31 Berthollet et al. (1787).
32 See the excellent study in Crosland (1962), section 3, especially chaps 6–8. Quotations are from pp. 198 and 179.
33 Martin Powell Davies, whom I supervised for his 'Using Historical Approaches to Improve Secondary Pupil's Image of Science' (1986), and for whose lively collaboration I am much indebted. I also thank Susan Leslie, of Brindley Hall School, Stantonbury Campus, Milton Keynes, whose classes were used in the trial scheme.
34 See, for example, Russell (1981).
35 Driver, Guesne, and Tiberghien (1985), p. 158.
36 Powell Davies (1986).

Practical Chemistry in a Historical Context

Peter Ellis

Introduction

In this paper I want to examine the role of historical teaching materials in today's secondary science courses, courses which are increasingly practical in nature. For the benefit of readers who may not be involved in secondary education I shall include a discussion of the place of practical work in the new GCSE courses. I shall present my own ideas regarding the use of historical material in practical chemistry lessons.

Many of my ideas were first developed for my thesis[1] written some years ago, but I hope this paper will stimulate discussion now and lead to some positive contribution to secondary science teaching.

I am conscious that my own knowledge of the history of science is sparse and superficial and that many of the ideas presented here have yet to be evaluated in a large scale test in school laboratories.

While the examples of lesson strategies that are discussed refer to chemistry, there is no reason at all why similar materials cannot be found to place physics and biology in a historical context. Similarly, while the discussion centres around work for GCSE the same strategies could be employed at pre-examination and advanced level.

History of Science in GCSE

For many years some teachers and academics have recognized that the history of science can be an important tool in the understanding of science and its importance to society,[2] although this view has not been accepted by the majority of teachers. In 1979, the writers of the ASE consultative document 'Alternatives for Science Education'[3] saw the history of science as part of one of the vital 'contexts' of scientific knowledge, and all the models of science teaching suggested in the document include a section involving a detailed historical approach to science teaching. Similar documents since have made similar references to including the history in science courses, culminating in the National Criteria for GCSE Science courses.[4] In the syllabuses produced by the GCSE examining boards, however, this ideal has been reduced to the requirement that students should be given ' . . . [an] appreciation of the developing and sometimes transitory nature of chemical knowledge, principles and models' (NEA Chemistry).[5]

The assessment objectives are even more vague as to the precise role of the history of science within the chemistry curriculum, but the requirement that students should understand the 'recognition of mistakes and misconceptions in chemical theory' is, perhaps, the most relevant. While the social, economic, environmental, and technological aspects of any science course is now expected to take up 15% of the syllabus and the exams, somehow the word 'history' has been left out. Yet it is a belief shared by many teachers that a historical slant is necessary to fulfil these aims and objectives in order to complete the scientific education of all pupils.

School science courses tend to present science as a fixed body of knowledge, correct unchallenged theories, and right answers, notwithstanding the new GCSE philosophy of teaching 'how science is done'. Investigation of some of the wild goose chases as well as the established route to accepted theories would surely give an insight into how science progresses. Similarly, the chemical industry is presented as a huge vital concern of great importance to our daily lives without consideration of how many different processes were developed over the years to cope with an expanding population with growing needs and changing aspirations. Pollution and environmental problems are often seen as a problem of the mid-twentieth century, whereas a study of eighteenth- and nineteenth-century industry would reveal that not all the problems are new and that similar problems have had to be faced in the past, albeit somewhat belatedly.

It is time that some of this was being done. Admittedly the Nuffield courses have over the years developed a number of historical topics,[6] although it has to be said that the historical option in the original Nuffield O-level course was the least popular,[7] and a number of the ASE's SATIS project units[8], designed to accompany GCSE courses, have a historical slant. There are few signs, however, that a historical viewpoint is looked on favourably, and no questions are likely to be set in GCSE exams which require a historical answer. There is, therefore, no obvious incentive for teachers to adopt material of a historical nature for their courses.

Nevertheless, there are a number of ways that historical material could and should be incorporated into GCSE courses to aid the understanding of chemistry.

1   The use of historical material to introduce topics, as used, for example, in the textbooks mentioned above, could be expanded to offer further examples to a more critical and informative level and thus give a broader and more balanced picture of the development of chemistry, instead of the bitty, disjointed titbits of information presented now. The disadvantage of this approach is that it just provides more information in courses that are already full and offers little assistance in developing pupils as chemists and scientists in general.

2   Some topics could be taught within an historical framework. Some teachers may already use this approach when introducing atomic structure; but the historical component is often restricted to a list of names and dates, with little information about the experiments that were carried out and reasons for theories being accepted or rejected. However at GCSE pupils are expected to be able to derive and test theories of their own from given data. Thus a teaching package describing the experiments that were carried out, the results obtained, and the interpretations contemporary scientists made from them, both correct and misguided, would not only make the topic clear to pupils but explain how science advanced. This approach could be applied to a number of topics of chemical theory, for example: the periodic table, elements and compounds, acid/base theory, and electrolysis/ionization. The principal problem to be overcome here is the lack of knowledge of teachers and their reluctance to try a teaching strategy that they feel unsure about. Thus full and complete teaching guides and classroom material would have to be provided.

3   Pupils are encouraged to investigate the chemical industry much more closely than in the past. This is relatively easy if the school is in an industrial area like Avonmouth, Teeside, or South Wales, and increased emphasis on the development of these industries would improve the pupils' awareness of their importance to the community. The South Wales ASE package on the growth of the iron and steel industry does this admirably.[9] In the past most parts of the country

had some kind of local chemical industry, such as limeburning, charcoal burning, metal smelting, brickmaking, dyemaking or soapmaking. An investigation of the chemical and technological aspects of these 'local' industries and their importance to the community, linked with field visits and industrial archaeology, would help pupils to realize the importance of the chemical industry in everybody's lives.

4　All the above suggestions are additional to present coursework and there may be difficulties in tacking on yet more material to already overburdened courses, however interesting and useful they may be. But there is one area of all chemistry courses central to GCSE philosophy which can be given a historical slant without significantly adding to the course content. That area is the practical coursework and assessments, and is examined more closely in the following paragraphs.

Practical work in GCSE

Chemistry is a practical subject with its facts and theories founded on experiment and observation. This has long been recognized by many teachers and, of course, the philosophy of the Nuffield courses has been the acquiring of knowledge through experiment. It is, however, only with the advent of GCSE that the skills of chemical experimentation have become an assessable part of the course. The National Criteria for science stipulate that at least 20% of marks should be awarded for internal assessment of coursework based on laboratory exercises. In addition, the bulk of the facts in the syllabus are intended to be drawn from experimental investigations. This has pushed practical work to the forefront of chemistry courses while the assessment of practical work has provided the biggest headache for teachers introducing GCSE.

The examining boards have divided the practical skills into four assessment abilities:

1　manipulation of apparatus;
2　observation and measuring and recording results;
3　interpretation of results and drawing conclusions;
4　designing and planning an experiment from a hypothesis and solving problems.

(Some boards have subdivided these abilities for assessment purposes.)

One difficulty teachers have is to make simple experiments, which may be of use for assessment, interesting to pupils who will probably have carried out similar procedures and techniques a number of times before. The intent is to teach good scientific practice; that is, the safe handling of materials and equipment, the principles of devising hypotheses and experiments, and interpreting results to arrive at acceptable conclusions.

I believe that a historical framework for practical exercises can be of significant use in the teaching and assessment of these practical abilities.

A Historical Context for Practical Exercises

My definition of the history of chemistry, which may differ from that accepted by professional historians of science, includes the work of recognized scientists, the development of scientific ideas, the development of the chemical industry (including metallurgical industries), and the use of chemical knowledge in crafts dating from prehistory. I used this definition when preparing a course entitled 'Alchemy to Chymistry' for use by top third year pupils.[10] The twelve period course included practical exercises as a central core of the course. The experiments were neither original nor complex and for the most part utilized materials and apparatus found in most chemistry laboratories. However the course provided a different context for performing these familiar experiments. The course was successful in that pupils were enthusiastic about the work but because of its length would be unsuitable for today's GCSE courses. Nevertheless, the success of the course encouraged me to link practical work with historical material in shorter units that can be used at any time in a chemistry course.

The historical material could be subdivided into three fairly arbitrary areas i.e.

1   the life and work of individual chemists;
2   the applications of chemistry (from craft based cottage industries to modern industrial complexes);
3   the development of chemical ideas.

In 1980 I devised six short teaching units, three in each of areas one and two. The three chemists investigated were Henry Cavendish, Humphry Davy and Michael Faraday, and the three industries the coal gas industry, the manufacture of sulphuric acid, and metal smelting.

Each of the units was designed to last one double lesson and all but the Davy unit included a practical exercise. The Davy unit included a demonstration of the preparation and properties of nitrous oxide. Apart from the written information and the practical exercise pupils also had to answer a few short questions based on the text and experimental results. At the time the pupils received the units enthusiastically and I have made use of them on a number of occasions since. It was easier to prepare the industrial units because dealing with the work of past chemists brought in problems of terminology, theories which have been rejected, and a need to place their work in the context of the time in which they lived. Also the work of later chemists is often too advanced and the experiments too complex to duplicate in the time available with the resources of a school laboratory.

Today all practical work should take into account the abilities which have to be taught and assessed at GCSE. Guidelines should be available to help teachers make their assessments, and the pupils' worksheets should contain all the necessary information so that pupils can work individually, leaving the teacher free to assess and provide assistance where necessary.

The experiments detailed in the six units discussed above lend themselves to the teaching and assessing of skills (i) (manipulation) and (ii) (observation and measurement). I have now prepared another unit in the industrial chemistry series incorporating these ideas which is included at the end of this paper. I hope teachers will examine and use it and comment on its usefulness. The unit examines the developments in the preparation and uses of salt and soda (sodium carbonate), two vital commodities which have moved from cottage industries to huge commercial concerns. This time I have included a number of practical exercises, some more difficult than others. Some or all of the exercises may be carried out as time allows, and one exercise may be carried out by one group of pupils being assessed while the rest of the class carry out the other exercises. Since many parts of the country have industries that in the past or present make or use these two commodities there is also opportunity for a visit or field work. For instance not far from my own school there is an area of marsh where salterns still exist, and it is possible to see the ponds where sea water at high tide was trapped and then allowed to evaporate.

The last area of the history of chemistry, the development of chemical ideas, did not figure in my earlier work although it provides opportunities for teaching and assessing the interpretation of data as a scientific skill and the postulation of hypotheses. One problem in trying to teach this skill is that pupils often fail to see how theories other than the current accepted theory could possibly be thought to be right. The theories of burning present a well known and relatively simple example of this. To most pupils the phlogiston theory is nonsense and incredible. This is because they have grown up with the oxygen theory which is (one hopes) common knowledge. Nevertheless the phlogiston theory does provide a good example of a hypothesis derived from limited data that was accepted for many years and was overthrown only by the development of new practical techniques in chemistry.

A unit has been prepared to examine this topic which can be used to help pupils learn how to make hypotheses and how to interpret data. The chemistry and the practical exercises are simple and the unit could conveniently fit into any topic on combustion or air. The unit requires pupils to make careful observations and to interpret this data to the exclusion of any other facts that they may previously have met in chemistry. It examines the importance of certain techniques such as the handling of gases and making careful weighings. An extension to the basic unit has been provided for quicker pupils which examines the danger of stretching a theory beyond its capabilities, in this case Lavoisier's incorrect definition of acids. Once again this unit has been included in this paper to allow teachers to use it and comment on it.

Next Steps

All eight of the short units mentioned above have been compiled into a booklet to which I have given the essential acronym CHiC (Chemistry in an Historical Context).[11] This booklet is available from the author, at least for the time being, for the cost of photocopying and postage (£1), and I would be happy for any teacher to copy the units and use them in school lessons. A synopsis of the CHiC course units is given in Appendix A, and the first unit, 'Henry Cavendish', is reproduced in Appendix B.

The main difficulty so far with this work has been the evaluation of the materials in an independent and objective way. Nevertheless lessons have been learnt and it is now time to think of strategies to ensure that such materials are incorporated into school courses.

First of all, the value of historical material must be promoted at all possible occasions, for example at teachers' meetings and conferences run by LEAs, examining boards, and of course the ASE, and through articles in magazines, newsletters, and bulletins.

Secondly the material must be available for teachers to use, cheaply and conveniently. Reference has already been made to the ASE's SATIS project, and this seems to me to be the ideal format – a wealth of ideas with teaching guides, pupil notes, and worksheets, attractively laid out but free of copywrite so that photocopies can be made as and when necessary. Above all it is cheap at about £3.50 per 10 units.

CHiC is on its way to being a SATIS type publication but there is some way to go. The units, particularly the first six, are only experimental and may well require revision in the light of GCSE. In addition I feel that many more contributions need to be worked through in order to produce a book that any teacher will be tempted to dip into. This is the important point; any publication must draw the attention of teachers who have not previously considered the history of science worth a thought or place in their teaching.

The units should follow a common pattern, as follows:

1   A comprehensive teachers' guide incorporating
    a   additional background notes;
    b   answers to questions;
    c   equipment and materials requirements;
    d   lesson plans and timings;
    e   assessment guidelines.
2   Pupil notes providing the context of the work and explaining the relevance and importance to modern chemistry.
3   A practical exercise, which may be a class experiment or teacher demonstration. (If it is the latter then assessment may be based on the interpretation of the results obtained.)
4   Questions, interspersed through the text, based on the background notes and on the practical work as well as extending the pupils' understanding of the topic.

The next step is to set up an editorial committee to evaluate contributions, plan the publication, and find a publisher willing to take it on.

APPENDIX A: CHiC – A SYNOPSIS

UNITS 1 – 3: 'BRITISH CHEMISTS OF THE EIGHTEENTH AND NINETEENTH CENTURIES'

1 *Henry Cavendish* Personality and brief biography; work on 'fixed air' and 'inflammable air'; faith in phlogiston.
Practical exercise: familiar experiments on the preparation and testing of carbon dioxide and hydrogen, and the cause of hardness in water.
2 *Humphry Davy* Early life and work on oxygen and respiration; work at the Pneumatics Institute on nitrous oxide and electrolysis; work at the Royal Institution on the safety lamp, discovery of sodium and potassium and the proof that chlorine was an element; later life and personality. Practical exercise: teacher demonstration of the principles of the safety lamp and the preparation and properties of nitrous oxide.
3 *Michael Faraday* Early life leading up to the post at the Royal Institution; working methods; laws of electrolysis and use of new names, anlaysis of steel, liquifaction of gases, analysis of coal tar, and preparation of optical glasses.
Practical exercise: investigation of the electrolysis of some solutions.

UNITS 4 – 7: 'DEVELOPMENTS IN CHEMICAL TECHNOLOGY IN THE EIGHTEENTH AND NINETEENTH CENTURIES'

4 *Coal Gas* Early experiments on firedamp, coal gas, and wood gas; first use in Paris for lighting and in balloons; Murdock and Boulton and Watt; first industrial use; Gas Light & Coke Co. development in London and elsewhere; improvements to process and to lamps; discovery and uses of components of coal tar – Perkin and Lister.
Practical exercise: the destructive distillation of coal.
5 *Sulphuric acid* Need for the acid in industry – textiles and the Leblanc Process. Early methods of preparation from green vitriol; Ward's process improved by Roebuck – the lead chamber process; improvements and pollution prevention; the contact process.
Practical exercise: Ward's method – burning of sulphur with sodium nitrate in glass vessels.
6 *Iron and steel* Importance of metals – noble metals, copper, tin, bronze; first smelting of iron – 'bloom' iron; development of the blast furnace; use of water power; limitations and problems of the use of charcoal; the Darbys of Coalbrookdale; smelting with coke, use of steam engines, the Iron Bridge; the Bessemer process.
Practical exercise: smelting of iron, copper, and lead oxides using charcoal.
7 *Salt and soda* Teachers' guide; need for salt; extraction from seawater; growth in demand hence faster methods.
Practical exercise: purification of rock salt. Need for soap; methods of manufacture; growth in demand for alkalis; potash; Leblanc process.
Practical exercise: second and third stages of the Leblanc process. Problems with the Leblanc Process – waste and pollution; improvements; Solvay ammonia process.
Practical exercise: preparation of sodium hydrogen carbonate by the Solvay process.

UNIT 8: 'A BURNING QUESTION – THE RISE AND FALL OF THE PHLOGISTON THEORY'
Teachers' guide; importance of combustion; the 'scientific method' – observation and theory;
Practical exercise: observation of burning wood. The phlogiston theory; Stahl.
Practical exercise: burning charcoal. Use of theory to predict results.
Practical exercise: test of the phlogiston theory – reaction of lead oxide with charcoal. Changes in practical techniques in the eighteenth century – collection and reaction of gases; inconsistencies of the phlogiston theory; Priestley; Lavoisier; the oxygen theory; Lavoisier's theory of acids, and its failure; Davy's proof that chlorine was an element.

APPENDIX B: BRITISH CHEMISTS OF THE EIGHTEENTH AND NINETEENTH CENTURIES

The following is the first unit in the CHiC course. Copies of all eight units are available from the author.

HENRY CAVENDISH 1731–1810

Little is known of the early life of Henry Cavendish since throughout his life he was a very shy man who shunned any kind of publicity. However it is known that he was born in Nice, sent to school in London and then to Cambridge University. After university he returned to London to live with his father, Lord Charles Cavendish, and worked in a laboratory set up in the stables. His father was interested in science and encouraged Henry in his investigations.

Henry was extremely shy and nervous in the company of others and rarely, if ever, appeared at social functions other than the regular meetings of the Royal Society. He rarely spoke but when persuaded to do so, spoke in a thin, shrill voice. Although a lord, Henry's father was not rich and Henry had to continue his experimentation with very few funds. However in 1780 Henry inherited a large sum of money but his only interest in it was that it freed him to get on with his investigations without financial worries. He rarely invited guests and when he did was not concerned with entertaining them. A leg of mutton would suffice for dinner. On one occasion five guests were expected and a maid plucked up the courage to ask him what he would like (Cavendish hated meeting his servants), 'a leg of mutton', was the impatient reply. When she suggested that a leg of mutton was insufficient for six perons he muttered, 'Well, two legs of mutton.'

In a period when science was expanding rapidly Cavendish familiarised himself with every bit of knowledge and made contributions to practically every branch of science. The chief characteristic of his work was that he took measurements in every experiment that he undertook, mere observation was not enough. His biographer wrote that 'His Theory of the Universe seems to be that it consisted of a multitude of objects which could be weighed, numbered and measured.'

Given the limitations of his equipment Cavendish performed his experiments with remarkable accuracy and although he published very little of his work because of his shyness, his genius was recognised by fellow members of the Royal Society.

Cavendish's first chemical work was with 'fixed air' discovered by Black in 1754. Like Black he prepared the gas from marble and hydrochloric acid and noted that the gas turned limewater milky. He went on to measure the solubility of the gas in water at different temperatures and also its relative density to air. At about the same time Cavendish was investigating the hardness of water in London and recognised the connection between the hardness and 'fixed air'.

Cavendish also prepared 'inflammable air' by the action of acid on zinc, iron or tin, and measured its relative density and observed the formation of water when it burned in air. Later he found that when metals, sulphur, phosphorus or inflammable air were burnt in air a fifth of the air was used up and no other gas took its place but when candles or oil were burnt 'fixed air' took the place of the air used up. With Priestley and Lavoisier's discovery of oxygen Cavendish's next experiment was to burn a measured amount of inflammable air with a measured amount of oxygen and measure the amount of water formed. He proved that all the gases used up became water and that water was formed by oxygen combining with 'inflammable air'.

However, all his life Cavendish refused to give up the Phlogiston Theory which stated that when a substance burnt it gave up phlogiston. By this theory oxygen was air without phlogiston and drew the phlogiston out of other materials i.e. made them burn easily. Cavendish thought that inflammable air was water combined with phlogiston and never accepted its real composition.

QUESTIONS
1 What are the modern names for 'fixed air' and 'inflammable air'?
2 Look at the diagrams of apparatus used by Henry Cavendish. What were they used for?

THE EXPERIMENTS OF HENRY CAVENDISH
1 Preparation of 'Fixed Air'
   Place a few marble chips in a test tube and add a little hydrochloric acid.
   Quickly fit a bung and delivery tube.
   Place the open end of the delivery tube in a beaker of water and collect the bubbles of gas in a test tube filled with water.
   Test the gas with a lighted splint.
   Now place the open end of the delivery tube in a test tube containing lime water (calcium hydroxide solution) and observe what happens.
2 Hardness of water
   Bubble the 'fixed air' into the limewater as in experiment 1 but continue bubbling the gas until the lime water goes clear again. You now have very hard water. Boil the solution in the test tube and observe what happens.
   Can you explain these reactions?
3 Inflammable air
   Place a few pieces of zinc in a test tube and add a little hydrochloric acid.
   Fit a delivery tube and collect the gas over water as in experiment 1.
   Try burning the gas when it is pure (i.e. a full test tube) and when mixed with air (i.e. remove the test tube from the water when it is about half full of gas and let air replace the remaining water.)
   Write down all your observations in your note book and explain what you have seen with equations.

NOTES

1 Ellis (1981).
2 Klopfer (1979), p. 87.
3 Association for Science Education (1979), pp. 37ff.
4 Department of Education and Science/Welsh Office (1985).
5 Northern Examining Association (1988), pp. 2–4.
6 Hunt (1979).
7 Haden (1977).
8 Association for Science Education (1986).
9 Brewer and Stone (1986).
10 Ellis (1977).
11 Ellis (1987).

Applying the History of Electricity in the Classroom
A Reconstruction of the Concept of 'Potential'

John Roche

Reviving the Arts of Explanation

There appears to be a growing interest among scientists and science teachers today in the historical dimensions of their subject. There have always been teachers, of course, who have recognized that a knowledge of the historical evolution of any topic gives a deeper understanding of it than that which is provided by textbook knowledge alone. What we are now witnessing, perhaps, is the beginnings of a general movement to integrate past and present in science as a means of consolidating and clarifying foundations.

My chief concern in this essay is to show that historical research by teachers can be a powerful tool in physics teaching, especially in the clarification, explanation, and exploration of physics concepts.[1] I shall take the concept of electric potential as a case history exemplifying this method.

Before I can begin it is necessary to defend this approach from an opinion current among some physics teachers and lecturers. According to this view false or made-up history can be an acceptable aid in teaching physics, much as Aesop's fables may be used to teach certain moral lessons. Aesop's fables, however, are indeed called fables, whereas the fabulous history of physics is usually presented to students as true. Historical fables in physics must surely falsify and deform scientific thought and method, since it misrepresents them. Accurate history of physics, on the other hand, lays bare the roots of modern concepts and illustrates the scientific mind as it really works.

This chapter is chiefly concerned with a particular historical development in physics which must strike anyone who compares Faraday's writings on physics with those of any modern physics textbook. Physics has become progressively more mathematical in the course of the past two centuries. As a result the arts of verbal explanation are now neglected in physics, or, at least, have not been developed to the same extent as mathematical and experimental presentations.

School teachers, teacher training colleges, and educational institutes have, of course, a strong professional commitment to clear verbal explanation. But the efforts of even the most brilliant teachers and lecturers are often hampered by the poor quality in physics today of the available techniques of qualitative definition, classification, description, drawing distinctions, and applying explanatory terms. These skills still thrive outside physics, of course, in biology, law, engineering, and in popular science, for example, but physics is particularly difficult to explain rigorously and clearly, and it requires the development of a broad and sophisticated resource of special explanatory concepts which can match adequately the complexity of the subject. I believe that such concepts can be developed, not simply by abstract reasoning, but also in the context of the effort to explain physics clearly in class. I also believe that this effort is assisted by a knowledge of the history of physics. I shall now develop each of these notions in more detail.

Every class in physics can be a fertile experiment in explanation. Physics teachers in the classroom learn to key their explanations to the explanatory expectations of the students. The students are already familiar with informal methods and categories of explanation drawn from everyday life and from many other disciplines. This is an extraordinarily rich resource which the physics teacher can draw upon and refine. It also has the enormous advantage that the student understands it and will understand physics explanations expressed in its terms.

It is common knowledge among physics teachers and educationalists that there are topics in physics which most students find difficult to understand. These include mechanics, electrostatics, electromagnetic induction, and thermodynamics. It is often assumed that the physics teacher who is bright enough and who labours sufficiently at preparation, presentation, and correction will be able to explain these difficult subjects clearly. This, I believe, is not always true. There are many topics, even in school physics, which no-one in the profession yet understands clearly. It is not that there is any missing evidence. It is because of the underdevelopment of verbal explanation, as I have suggested. Physics, I believe, has not yet recognized the importance of creating an adequate repertoire of carefully selected and uncontroversial explanatory concepts and terms which can be used to explain difficult topics clearly. There is, of course, a considerable body of successful explanatory concepts already available in physics, but this needs to be improved upon and greatly extended. Reformation of this sort, rather than some new revolution is, I believe, needed in physics to-day.

To my knowledge there are no research institutes, staffed by physicists, which are devoted exclusively to clarifying, explaining, and understanding the foundations of established physics. All able physicists are busy pushing back the increasingly remote frontiers of science. Understanding physics seems to be low-priority knowledge which one picks up casually while studying, teaching, supervising laboratory work or carrying out one's own research. Nevertheless, most would admit that clear understanding is the key to progress in physics, and the lack of understanding is the cause of much intellectual distress and alienation from physics. Furthermore, understanding a difficult concept is one of the most satisfying intellectual experiences known to man.

History, I believe, is essential to any endeavour to improve the quality of understanding and explanation in physics. Many, if not most, of the difficult topics in physics teaching can be made understandable only by joining a thorough familiarity with the modern theory to research into their historical origins. There seem to be two main reasons for this. Firstly, history lays bare the roots of the modern theory and this often reveals that misinterpretations, errors, and obscurities have worked their way into the very foundations of the theory and have never been cleared up subsequently. Secondly, a difficult topic, seen through the prism of history, often turns out to be a nexus of competing traditions and conceals a variety of distinct problems. In other words, concepts which should be carefully distinguished sometimes posture as a single concept. When all of this is understood in detail many difficulties become much easier to resolve.

The Prism of History

After these general remarks some concrete examples will help to make my meaning plain.

One kind of difficulty inherited from history is obsolete technical language. This is only a problem, of course, when the individual teacher, or the physics profession as a whole, fails to recognize a certain term as redundant.

The 'permittivity of a vacuum', implying as it does an electrical property of the vacuum or free space, was introduced late last century by Oliver Heaviside (1850–1925),[2] because it was widely believed that space is filled with a material aether which permits the passage of the electric field to a definite degree.[3] Today, the physical constant so named is known to be the 'electric interaction constant'[4] between charges, analogous to the 'gravitational interaction constant' between masses. The term 'permittivity of a vacuum' is, therefore, obsolete. Nevertheless, it lingers on. Other examples of obsolete terms are 'electromotive force', which is usually used to mean a voltage rather than a force, 'magnetic coercive force', which is a magnetic field strength, and the 'permeability of free space'.

Another closely related source of explanatory difficulties may be due to a lack of awareness that there are remnants of falsified theories which still survive in various odd corners of modern textbooks and research literature. Since the last century there have been at least nine competing theories of the electromagnetic field.[5] Only one of those is generally accepted to-day, but remnants of the others, often unrecognized for what they are, still struggle to survive on the margins of modern electromagnetism. I am not criticizing their right to survive, but I am criticizing mixed or incoherent explanations which result from a lack of historical knowledge of the evolution of electromagnetism. Falsified theories may still be useful, of course, as conventional or fictional explanations. For example, the use of magnetic poles is an easier way to 'explain' many magnetic phenomena than the true force between electric currents. Such 'explanations' may still be used harmlessly, provided their fictional or conventional character is made quite explicit. In many cases, however, this is not brought out.

A good example of this is the so-called catapult field in electromagnetism. It is not difficult to imagine the perplexity generated by teaching in one class, correctly, that the applied magnetic field exerts a force directly on the moving electrons in the wire, and in the next class teaching, incorrectly, that the force on the conductor is due to an interaction between the applied magnetic field and that due to the moving electrons in the wire. In modern macroscopic electromagnetism there is no interaction between magnetic fields themselves, and no tension in the lines of force. Fields act only on charges, not on each other.

To consider another example, it is not yet clearly understood in physics, I believe, that the force between two bar magnets is *not* composed of forces between the poles. In fact, each part of a magnet contributes a roughly equal amount to the total force. This total force can indeed be notionally resolved into components which act between the poles, but the true force is a distributed force.

Following Faraday it is often said that an iron sphere 'shields' its interior from an external magnetic field. This, according to strict modern theory, is incorrect. The applied magnetic field passes right through the iron undiminished. It then induces a secondary magnetic field in the iron. The resultant of these two fields gives rise to a balanced field in the interior. It is best, perhaps, to call this process 'effective' shielding.

No teacher or lecturer or textbook writer deserves criticism for making any of these mistakes – they are quite endemic and are usually unrecognized within the profession. Historical research lays bare the origins of these difficulties and goes a long way towards clearing them up. Problems of this sort cannot be resolved by long hours of pondering over modern texts. This can be amateurish and misleading because it is most likely that the very concepts which need to be cleared up have long been internalized and, without critical history, will infect and obscure the effort to analyse them. He who is not part of the historical analysis is almost inevitably confined to the received synthesis.

Origins of the Concept of Potential

As a case history illustrating the method of research which I call 'critical physics', I shall consider the difficult concept of electric potential or voltage. The physics teacher or lecturer is prompted to mull over this concept year after year in the course of class preparation and delivery. If he or she is also provided with a knowledge of the evolution of the concept then there is an enviable opportunity to treat each class on the topic as a workshop in which new or modified explanations of potential can be proposed and tested against the common sense understanding of the students. This should lead to a much improved explanation of potential, one which gathers together all of the valid insights of the past, goes beyond them where necessary and fashions a more comprehensive, unified, and coherent concept. In the context of the classroom itself, therefore, the schoolteacher or tutor has the opportunity to carry out important foundational research in the concepts of physics.

How can historical research prepare the ground for such an improved concept of electric potential? A sketch of the evolution of the concept should help to make this clear.

The modern concept of potential is the fusion of at least five quite distinct historical traditions. Despite the seeming unity of the received concept, each of these traditions still plays a semi-autonomous role in the present-day understanding of potential.

One of the earliest hypotheses concerning the nature of potential was that of William Watson (1719–1787), who proposed the view in the late 1740s that the electric fluid is in a state of compression or pressure. Henry Cavendish (1731–1810) developed this view further in the 1780s.[6]

The concept of potential as a pressure can still be found in textbooks and other literature today, sometimes as an analogy and at other times as a literal explanation. Although this explanatory tradition is now marginal it illustrates how a tradition of explanation, once launched, seems to remain in service for ever.

In a publication of 1779 Allesandro Volta (1745–1827) introduced his powerful concept of 'the degree of electric *tension*' of a charged conductor.[7] He showed how this 'tension' could be measured and he transformed it into a successful working concept. It was immensely influential. It even gave rise to Faraday's concept of tension along the lines of force, and influenced Maxwell's theory of stresses in the aether. Volta's concept of tension survives in electrical engineering where expressions such as 'high-tension lines' are still used.

Volta also introduced the concept of 'electromotive force' as the prime mover of a current in a closed circuit, and measured it in terms of his electric tension.[8] G.S. Ohm (1789–1854) in 1827 attributed the concept of 'electroscopic force', or tension, to circuit resistors, a concept now known as the voltage across the resistor.[9]

It should be noted that these early concepts drew heavily on mechanical analogies. Furthermore, although electric 'pressure' or 'tension' was thought of as a property of the electric fluid in a conductor, it was not thought of as a property exercised across the neighbouring space. This extension of the concept was added later in the nineteenth century.

A very different concept, that of electric 'potential', was introduced in the early nineteenth century not by experimentalists, such as Watson and Volta, but by the mathematical physicists Poisson and Green. S.D. Poisson (1781–1840), in 1811, strongly inspired by Laplace, introduced a mathematical function to electrostatics whose gradient was numerically equal to the local electric intensity, or force per unit charge.[10] This was a very convenient function because it was non-directional and easier to handle mathematically than the electric intensity function. It is important to note that the potential function was introduced to electrostatics as a mathematical artefact only, and not as a physical state.

George Green (1793–1841) of Nottingham, in 1828, invented the term 'potential' for the potential function and developed its mathematical properties much further.[11] The potential function had a fixed value on the surface and in the interior of an electrostatically charged conductor, but it also had values in space everywhere in the neighbourhood of that conductor.

J.L. Lagrange (1736–1813) and C.F. Gauss (1777–1855) introduced and developed the mutual potential energy function, which is related to the potential function. This function was also thought of at first as a mathematical construct only.[12]

The next major step in the evolution of the concept of potential was made by Gustav Kirchhoff (1824–87), who effected a synthesis in 1849 which showed that Volta's tension and Poisson's potential function were numerically identical in a conductor and that they should, therefore, be reduced to a single concept. He also associated the new concept with electrical energy.[13]

The concept of potential does not seem to have developed very far beyond the point at which Kirchhoff left it in the middle of the nineteenth century. There is now a cluster of at least four concepts which are closely related to each other and yet are thought of as somehow distinct, namely potential, e.m.f., circuit voltage, and electrical potential energy. Engineers and teachers are quite confident that voltage is a genuine (and dangerous) physical property, while some theoretical physicists still suppose with Poisson and Green that potential is a mathematical artefact only. All of this suggests that a considerable effort is now required to distinguish, clarify, and formulate a coherent theory of potential.

Suggestions for the Clarification of the Concept of Potential

Introduction

The argument which follows attempts to be systematic, since it might be tedious for the reader were I to try to describe the steps by which

I arrived at this interpretation of potential. Suffice it to say that I have applied the methods outlined earlier in this article, namely historical study, the analysis of concepts, and trials in class. Although this effort has gone on for almost thirty years I feel that there is still much to be done. Indeed, I believe that the work of clarifying and understanding the concept of potential, and of all other concepts in physics, will never end.

Potential energy

There appear to be at least two kinds of electrical potential energy, namely mutual potential energy and individual potential energy. The *mutual potential energy* of a system of charges may be defined as the total energy available in that system to supply to the charges composing it, while these are being withdrawn to a great distance from each other. It may be positive or expulsive, or it may be negative or binding. The mutual potential energy of a system of charges is zero when all charges are electrically remote from each other.

A more frequently encountered type of potential energy in physics is the potential energy of a single charge with respect to a fixed system of charges. The *individual potential energy* of that charge is the work which the system of charges will do on it, while it is being fully withdrawn from the system, the other charges remaining fixed. The energy given to a proton released near the surface of a positively charged Van De Graaf generator, and driven far away, is a positive example of individual potential energy. The ionization or escape energy required by an atomic electron is an example of negative individual potential energy.

Individual potential energy is usually called *absolute potential energy* in order to distinguish it from *relative potential energy*. The latter concept is familiar in elementary gravitational theory where the relative gravitational potential energy of a given body with respect to an arbitrary reference level is the work which the system is ready to do on the body as it is transferred from its starting point to the reference level. Relative electrical potential energy is similarly defined. It is used, for example, in discussions of excitation and de-excitation of electrons in atoms.

The absolute potential energy of a given charge with respect to a system of charges is a true physical state, since a system loses rest mass when a charge with expulsive potential energy is removed, and gains rest mass when a bound charge is withdrawn.

A charge has zero absolute potential energy with respect to a given system of charges when it is far away from that system. This follows

from the consideration that no work will be done by the source charge
on the test charge as it is further withdrawn, and absolute potential
energy is defined precisely in terms of the latter process. Although it
has no absolute potential energy a remote charge has an infinity of
relative potential energies with respect to points in the neighbourhood
of the source charge.

Electric potential

Electric potential is explained and defined in terms of individual rather
than mutual potential energy. When a test charge is withdrawn far
from a system of fixed charges the work done per unit positive charge
is independent of the magnitude of the charge. The electric potential is
defined as this invariant property. If the potential at a point ϕ, defined
in this manner, is called the absolute potential there, then it is related
to the absolute potential energy Φ of a charge placed there by the
equation

$$\Phi = Q\phi$$

Clearly, the absolute potential at a point is the absolute potential energy
per unit positive charge placed there. The work W done on a charge
as it leaves or enters the system is expressed by $W = +Q\phi$ and $W
= -Q\phi$, respectively.

By analogy with the expression 'intensity of electric force', which is
the general meaning of 'force per unit charge', I shall describe the
electric potential in general, at a given point, as 'the intensity of the
potential energy of charges placed there'. When we think of potential
at a point, therefore, in this interpretation, we must suppose some
charge there in possession of that property. If no such charge is present
then potential is a virtual rather than an actual property at that point.

Together with an explanatory definition of potential a measuring
definition is, of course, required in order to make the concept physically
complete. The present standard measuring definition of potential at a
point is the work done by an agent against electrical forces, per unit
positive charge, in transferring a test charge from infinity to the point
in question.

The term 'infinity' is introduced here for mathematical convenience
only, since, in an actual measurement, it is sufficient for the starting
point to be far enough away from the source of the potential so that
the electrical effects of the source are negligible.

If the measuring convention were specified in terms of the electrical work done by the source system itself, per unit positive charge, on a test charge which exits *from* the system, this would be in better agreement with the physical concept of potential as I have explained it above. There are several other advantages as well. As a working concept in physics potential is used where the electric forces themselves do the work, not some notional agent. The received definition, therefore, displaces attention from the physics of the situation. Next, the new definition leads to a more natural algorithm for the potential function. Finally, the new conception, unlike the old is not likely to be confused with the concept of relative potential energy discussed above. Both conventions, of course, give the same magnitude and the same sign to the potential.

Whatever the exit path chosen for a test charge from a given point in the neighbourhood of a system of stationary charges, the intensity of the electrical work done on it will be the same. Furthermore, the total electrical work done in any closed path is zero. Systems of force with this property are called conservative. Since possible exit paths radiate from a given point in all directions, and since the terminal point chosen is arbitrary, absolute potential has no specifiable direction. It is, of course, unique and single-valued and zero at points electrically remote from all charges. Furthermore, it is positive or negative depending on the sign of the source charge.

The absolute potential at any point which is electrically remote from a given system of charges is zero with respect to that system, since the absolute potential energy of charges placed there is zero. The absolute potential at a point which is electrically remote from *all* charges will, of course, be absolutely zero. Those who suppose that the absolute potential at a point in empty space can be given arbitrary values are clearly treating it as a mathematical artefact only, and are carrying to an extreme the purely mathematical interpretation of Laplace, Poisson, and Green discussed above. However, when interpreted as the intensity of exit energy, absolute potential has a well-defined physical meaning which is analogous to the specific latent heat of sublimation, for example. Auxiliary non-physical potential functions can, of course, be very useful, but only the absolute potential deserves to be included in the physically descriptive equations of electromagnetism. In order to distinguish auxiliary from physical potential functions, it is helpful to think of the latter in relation to the charge system which causes it, and not abstractly.

Mathematically speaking, the expression

$$\phi = \frac{1}{4\pi\epsilon_0}\frac{Q}{r}$$

associates a definite numerical value of the potential function ϕ with every point r in the neighbourhood of the source charge Q. Physically speaking, however, this can be misleading. From a physical point of view potential is a commitment of the system to supply energy along a stellation of exit paths from the given point. It is not, therefore, a property *localized* at the latter point, such as the electric force on a charge placed there.

Some properties of the electric potential

The *potential difference* between two points is the difference in absolute potential between them. This property is inadequately defined unless the starting point A and the terminal point B are specified. The potential difference is then the final value less the initial value,

$$\Delta\phi = \phi_B - \phi_A$$

Absolute potential decreases most rapidly in the direction of the field. Respecting direction, it is easy to show that the electric intensity is equal to the negative potential gradient,

$$E = -\ \mathrm{grad}\ \phi$$

Adding a positive charge anywhere in space increases the absolute potential everywhere in its neighbourhood, because its presence creates an additional commitment to supply energy to every exit path radiating from every neighbouring point. The addition of a negative charge, or the removal of a positive charge, similarly reduces the neighbouring potential everywhere.

An uncharged insulator placed in the neighbourhood of a positive charge acquires the potential previously possessed by that region of space which it now occupies.

A charged conductor which is placed in the vicinity of another charge will have a resultant potential which is partly due to its own charge and partly due to the neighbouring charge. The interior of the conductor will also have a potential since any path from its interior to any electrically remote point will be an electrical energy supply route.

When a conductor of higher potential is joined to a conductor of lower potential, positive (conventional) charge must flow from the former to the latter because a mean negative potential gradient implies an average positive electric field, which will drive the conduction charges in the connecting wire.

Any charges on the Earth's surface, or in the atmosphere locally, will contribute to the absolute potential of a given conductor or insulator. Instead of dealing with the absolute potential of a conductor it is often more convenient to deal with its potential relative to the local Earth's surface.

Each plate of a charged capacitor will have an absolute potential which is partly due to its own charge, and partly due to the charge on the opposite plate. The potential difference between the plates is simply the difference in these absolute potentials.

When an uncharged dielectric is inserted between the plates of an isolated capacitor the charges on the plates cause a displacement of positive electricity in each atom towards the negative plate. This increases the absolute potential of the negative plate. Similarly negative atomic charge is displaced towards the positive plate, reducing its absolute potential. The net effect is to reduce the potential difference between the plates. When either plate of a charged capacitor is joined to the Earth, this causes a small flow of charge into or out of that plate which equalizes its absolute potential with Earth potential. This shift of charge also alters the absolute potential of the other plate by an approximately equal amount.

It should be clear from all of this that absolute potential can only be understood clearly when it is closely linked in thought to the charge which causes it.

Voltage

Electric potential is produced by unbalanced charge systems only. Electrically neutral bodies can produce forces which do work on charges even in the absence of absolute potentials or potential differences. When a magnet is moved along the axis of a closed circular conductor, for example, the induced electric field of the magnet drives a current around the circle without the presence anywhere of potentials, potential differences, or unbalanced charges.

There is a need, therefore, especially in circuit theory, for a broader concept of the intensity of electrical work which is able to encompass all cases in which work is down on charges. In circuit theory also, it is necessary to be able to specify the sense or direction in which work is being done. The concept of voltage satisfies both of these requirements.

The magnitude of the voltage between two points is the work done per unit positive charge passing between the points. The sense or direction of the voltage is that of the field or conventional current passing between the points. Sense or direction has many meanings in physics and the present meaning does not imply that voltage is a vector quantity. Like the current in a circuit, voltage is a pseudo-scalar, that is, a scalar with a sense. Voltage is always specified, therefore, between two points or bodies. There is no absolute voltage at a point.

In a resistor carrying a current, for example, the sense or direction of the voltage is the direction of the conventional current, or of the driving electric field, since that is the direction in which positive work is being done on the conventional current. In a charged capacitor the direction of the voltage is from the positive to the negative plate, *through the capacitor itself*. The direction of an e.m.f. is the direction in which it freely drives the conventional current *through itself*.

When a charge of arbitrary sign passes in any direction between two points which have a voltage V between them, the work done by electrical forces on the charge is given by $W = \pm QV$. The discussion above makes clear which sign should be chosen.

Electromotive force

Electromotive force has two meanings in physics. It means the prime mover of the electric current in a circuit, and it also means the voltage across that prime mover. I shall reserve the abbreviation 'e.m.f.' for the latter meaning only.

The most important cases of electromotive forces for the present study are those operating in voltaic cells and in electromagnetic induction. In a cell the prime mover of the current in the circuit is a very short-range electrochemical force-field which is localized in the contact zones between electrode and electrolyte. Induced currents are primarily driven by more extensive induced electric fields, or by magnetic fields, which are, nevertheless, usually confined to only a small portion of the circuit. Electromotive forces are non-conservative. This means that they can supply energy to charges which move in a closed circuit, unlike electrostatic force fields.

Electromotive forces, although they may be localized in a small part of a circuit only, nevertheless drive a uniform current throughout the circuit. This happens as follows. The charges which are driven out of the active zone, in the first instance, become excess charges and drive onwards those conduction charges which are ahead of them. A corresponding process of attraction occurs at the negative end of the active zone, and in succession all around the circuit, causing a current to flow everywhere. At each point of the circuit the excess charges there repel each other and travel immediately to the surface of the conductor. As a result a surface charge gradient is established all around the inactive zone, ranging from positive at the positive terminal of the e.m.f. to negative at the negative terminal. These surface charges set up an electric field which is the driving agency of the current in this part of the circuit. The auxiliary charge distribution itself is maintained and energised by the e.m.f.

In the zone occupied by the e.m.f. there will, of course, be a reversal of charge gradient, which is gradual for an induced e.m.f., but quite sudden in the contact region between the poles and electrolyte of a cell. This means that there will be a reversed electric field in these zones which is in opposition to the local e.m.f., and which ensures, among other things, that the current there is no greater than that in the rest of the circuit.

Since the fields produced by the auxiliary surface charges are conservative, it follows that the voltage across these fields outside the region of the circuit containing the e.m.f. must be exactly equal and opposite to the voltage across the corresponding conservative or coulomb fields within the region occupied by the e.m.f.

If the direction of the e.m.f. is chosen as the positive sense of circulation, then the net internal voltage will be the difference between the e.m.f. and the coulomb voltage. It follows, therefore, that the algebraic sum of the net internal voltage and the external circuit voltage will be exactly equal to the e.m.f. This means that all of the energy supplied to a charge as it passes once around the circuit comes ultimately from the e.m.f.

In a voltaic cell the net voltage between the poles measured internally is simply the voltage across the electrolyte resistance, since the coulomb voltage virtually balances the e.m.f.'s in the contact zones. On open circuit, this net internal voltage is zero and the external coulomb voltage is exactly equal in magnitude and circulatory sense to the voltage across the internal electromotive force. When the circuit is closed, there is a fall in external voltage which is exactly balanced by a rise in the voltage across the electrolyte.

In a transformer or inductor winding, the net internal voltage is again the voltage across the internal resistance only.

If a unit charge is taken once around a circuit the total work done on it by the resultant fields is the sum of the net voltages in each circuit element, which may include cells, capacitors, resistors, coils, and even air-gaps. But the work done on the charge in this closed circuit by the purely conservative fields present must be zero. These therefore drop out of the sum and only the non-conservative e.m.f.s remain. Hence Kirchhoff's second network law follows: that the sum of the underlying e.m.f.s in any closed circuit is exactly equal to the sum of the net voltages in exactly the same circuit.

Absolute potentials in a current-bearing circuit

In describing electric circuits it is clearly important to distinguish carefully between two interacting structures, namely, the auxiliary surface charge system which is conservative and which creates an absolute potential at every point of the circuit, and the electromotive forces which are not conservative and which do not contribute to the absolute potentials.

It is also necessary to distinguish between the absolute potentials and voltages in a circuit. The absolute potential of any point of a circuit is essentially an external property of the circuit, since it relates to exit energy. Furthermore, it is almost always a latent or inactive property. Voltages, on the other hand, are internal properties of circuit elements and measure active energy processes. This means that when internal processes only in a circuit are being discussed, there may be no need to introduce absolute potentials or even potential differences. For example, only e.m.f.'s and voltages appear in the above statement of Kirchhoff's law.

When separate circuits, or separate branches of a single circuit are being compared, however, it may be necessary to introduce potentials. For example, the action of a potentiometer can be explained by stating that the galvanometer current is zero because the absolute potential of the slider is the same as that of the point of the potentiometer wire where it makes contact.

It is also important to note that a current-bearing circuit can be electrostatically charged, just as any conductor can be charged, but there are peculiarities. Since the absolute potential varies from point to point around a circuit, the amount of charge which flows into the circuit from a given charged conductor will depend on the potential of the point of contact chosen and on the potential of the charged conductor.

If, for example, any point of a circuit is joined to the Earth, charges will flow to reduce the potential of that point to earth potential. These charges will spread over the whole circuit and the inflow (or outflow) will cease only when the absolute potential of every point of the circuit has been reduced by an equal amount. Of course, only one point of the circuit will be at earth potential, and only the absolute potentials and the surface charges and fields are affected by earthing, not the circuit voltages, nor the differences in absolute potentials. Earthing a circuit protects the absolute potentials of a circuit from random fluctuations caused by local movements of charge and by accidental contacts.

Sign conventions and terminology

When Kirchhoff's second network law is applied to a circuit, two sets of sign conventions must be introduced in sequence to each of the circuit elements. In the first stage the e.m.f.s and voltages are given their proper directions, if known, or, if not, are assigned tentative directions. In the second stage a clockwise or counterclockwise circulation convention is chosen, for convenience, for each sub-network. An additional positive or negative sign is then attached to each e.m.f. or voltage symbol, depending upon its direction in relation to the circulation convention. As in geometrical optics, these conventions, if applied consistently, allow a single general formula to be applied correctly to all cases.

For example, when a cell is applied to a resistor, the voltage across the pure electromotive force of the cell, that is, the 'e.m.f.' of the cell, is in the same direction as the voltage across the resistor, when measured with respect to the circulation convention in the circuit. Not only does the resistor voltage not 'oppose' the e.m.f., it is the agent of the latter in supplying energy to the current.

When a capacitor discharges through a resistor, the voltage across the capacitor and that across the resistor are in opposite directions, again with respect to the circulation convention in the circuit. This is so because, in such a circuit, which contains no e.m.f., all fields are conservative and the net voltage around any closed circuit must be zero.

It is sometimes stated that a capacitor charges up until its voltage 'balances' the e.m.f. Here, the capacitor voltage and the e.m.f. act in the same circulatory sense and they act through different bodies. It follows, therefore, that they cannot balance each other. However, the electrostatic voltage between the charges on the battery terminals measured *through the battery* does indeed balance the e.m.f., when the capacitor is fully charged.

Conclusion

I would recommend a sharpening of the conceptual, verbal, and notational distinctions which are already made, although not very consistently, in this subject by physicists. I believe that it would clarify explanations and practical usage were the term 'potential' and the symbol ϕ reserved for 'absolute potential' or the difference between two absolute potentials, only. 'Voltage' and the symbol V seem best reserved for the concept, already established in circuit theory, of the directed work done by local electrical forces, per unit charge, along paths between two designated points. This means that we may correctly speak of the 'voltage across' an electromotive force, or the 'voltage between' capacitor plates, but it means that expressions such as 'voltage difference' or 'voltage drop' become as meaningless as 'work difference' or 'work drop'. It also implies, for example, that the 'voltage of a conductor' and the 'potential drop across an e.m.f.' are both incoherent expressions.

I would finally recommend that 'electromotive force', understood as the prime mover of electric current, be carefully distinguished from 'electromotive voltage (e.m.v.)', understood as the voltage across that prime mover.

NOTES

1 Roche (1987).
2 Heaviside (1892).
3 Everett (1901), p. 57. '... It is called the permittivity ... because it corresponds ... to the facility with which the elastic membranes yields to the force which displaces the liquid'.
4 *Concise science dictionary* (1984).
5 Whittaker (1961, 1962).
6 Watson (1747), Cavendish (1879), Heilbron (1979).
7 Volta (1779), p. 262.
8 Roche (1987).
9 Ohm (1841), p. 403.
10 Poisson (1812).
11 Green (1828).
12 Lagrange (1867–92), Gauss (1841), p. 176.
13 Kirchhoff (1883, 1850).

Play Up and Play the Game

A Simulation of Hard and Soft Fraud in Science

Edward Yoxen

The Value of the History of Science

Many working scientists view the history of science as a kind of entertainment, for which they will occasionally make time. For them it is slightly less important than watching horror movies and slightly more elevating than a visit to the pantomime at Christmas. The justification is either that the important achievements within one's field are so well known to any competent practitioner that there is no particular need to study them in more detail, or that the past is so littered with errors and half-baked ideas that it is pointless to expend any effort only to discover how frequently one's predecessors went wrong. At best the genesis of the central ideas of one's chosen field can be studied in the peace of one's retirement, when competitive pressures can at last be ignored.

Similarly it is often argued that students at school or university should have no exposure to the history, philosophy, or sociology of science, since the development of a technical competence must be their primary goal. For this purpose only the absorption of current facts and concepts, the performance of well-tried experiments that illustrate principles still in use, and the solving of exemplary problems from the very recent past will do. Given the stress of a modern research career, this view of historical study is understandable, but it is surely wrong, for a series of reasons.

Firstly the past is never self-evident from an examination of the present. Although scientific knowledge usually develops by accretion, much of the reasoning needed to attain any given stage is swiftly lost from view. The construction of scientific papers alone sees to that.[1] Only the most superficial and partial knowledge of the history of science is derived from scientific training, except perhaps in those fields, and embryology is one, where enduring conceptual difficulties make a serious engagement with earlier work heuristically useful.

Secondly it is surely rather arrogant to suppose that one has nothing to learn from chains of reasoning carried out 100, 200, or even 2000 years ago. Even 'mistaken' or 'outmoded' representations of nature often turn out on historical examination to be remarkably subtle and well-constructed, whatever their predictive power or consistency with evidence available to us now. Thirdly history reveals the complexity of method. Students are taught that 'the scientific method' involves the formulation of a hypothesis, from which predictions can be derived, which are then tested empirically in experiments, leading to the derivation of new consequences or the reformulation of the initial hypothesis. Case-studies reveal however a bewildering variety of research strategies and methodological conventions, some of which fit this simple picture, but most of which do not. Over the last 50 years philosophers of science have quarried the past for material with which to build much more sophisticated models of the research process. Whereas they used to look to history for edifying cases ripe for idealization, nowadays they are much more sanguine and open-minded about the inconsistencies they find.

Fourthly the historical approach lends perspective to practices, conventions, forms of social interaction, and professional institutions which may seem natural and inevitable, but which have evolved over time, as the environment within which science is practised has changed. For example we now think of competition as an inevitable part of science and of publication as a way of demonstrating originality under competitive conditions, but the rules that indicate what is acceptable to do to cope with the fact of competition or which regulate how one claims to have made an original contribution are themselves historical products.[2]

Or we may suppose controversies arise in science simply because of uncertainty about the available evidence or how to interpret it. Rarely is this the whole story however, and a great deal of historical scholarship over the last 25 years suggests that scientific debates are conditioned by all kinds of cognitive, psychological, institutional, and cultural factors that are inextricably linked with technical and theoretical questions. This is not to say that such a view of science, and the related question of what one means by objectivity and rationality, is universally accepted, but one may argue nonetheless that the implications of the competing arguments about the nature of science are so important and their evaluation so educative, that to exclude them from the scientific curriculum is very irresponsible. People who have no comprehension of the ground-rules of their craft, or who have no idea how to make an intellectually coherent defence of them, are quite likely to be dogmatic, arrogant, and philistine. Given the power and pervasiveness of science today that represents a problem for us all, whether we be scientists or not.

Finally there is the question of self-knowledge. In being formed as scientists, if only in a preliminary way, people not only acquire technical understanding, they also learn the value of certain emotions and the irrelevance of others. Science is largely about the manipulation of abstract models of the world or actual material systems, that are treated as disconnected from the world as a whole. In an important way scientific socialization is a process of learning to be comfortable with certain forms of abstraction, of coming to think of oneself as a dispassionate explorer of those facets of the material world that can be revealed by experimental or theoretical analysis. It is also about learning to accept or even enjoy highly competitive social interactions, throughout one's working life, in the hope of uncovering some very general principle or mechanism.

For example, James Watson's best-selling account of his own early experiences in biology, *The Double Helix*, was written as a fable for young scientists, to inspire them with the idea that success in research today is granted only to those who have prepared themselves intellectually and emotionally for a life of endless competition.[3] The book is also a work of self-justification, as Watson reveals how single-minded he and Crick were in picking up clues from others' work that had apparently been missed by the original investigators. It exults in the satisfactions of a certain rather risky research mode.

Manifestly then people need to understand and come to terms with the feelings and dispositions than an involvement with science will generate. Here educational institutions have a role to play. In this spirit then this paper deals with fraud in science, as an issue where historical studies have something to teach us about the scientific enterprise. In particular they reveal that historical analysis is not an elementary or trivial exercise, and that, just as is the case in science, its results can be both definitive and open-ended.

Hard and Soft Fraud in Research

In recent years a number of rather spectacular instances of scientific fraud have come to light, particularly in the United States. A young Harvard cardiologist, John Darsee, was discovered to have made up the data for most of his major research papers, some of them co-authored with very distinguished researchers. A brilliant young molecular biologist, Mark Spector, claimed, again with the backing of an internationally renowned senior colleague, to have worked out a general theory of the changes in cells that lead to cancer. Eventually sceptical colleagues discovered that the results, on which everything rested, were an ingenious forgery. In Australia, a well-respected biochemist, Michael Briggs, Dean of the Science Faculty in his university and a consultant to the World Health Organisation, was shown to have invented the data in a whole series of collaborative studies in steroid biochemistry, of major importance in assessing the safety of new types of oral contraceptive. In all these cases there are major problems in deciding which of their results, if any, can still be trusted and quite how to judge people who should have suspected what was going on. They reveal, contrary to the view often promulgated about science, how much is taken on trust, and how selective the processes of criticism and appraisal often are.

These are all cases where the intent to deceive has been admitted, or is beyond dispute. The term fraud, with a technical meaning in law and connotations of intentional deceit in business transactions, is best reserved for cases of this kind. The list of well-documented and suspected cases of such flagrant research malpractice is now extensive.[4] Research funding agencies and universities have been forced to develop procedures, which hitherto they often lacked, for dealing with such problems, and the professional literature now contains hundreds of commentaries and position papers on the subject.[5]

However it is clear that methodological misdemeanours form a continuum, extending from the complete fabrication of one's material, sometimes for years on end, to relatively trivial exaggerations and plagiarism. Whilst the harm done by wholesale fabrication is obvious, it is certainly arguable whether, at the other end of the scale, failure to observe the niceties is harmful at all. Indeed some people would argue that a certain carelessness about evidence and its authorship is a price well worth paying for progress in research.

It is also clear that many of the major figures in the history of science, Newton and Mendel among them, doctored their results to make them seem more plausible. For the most part their reputation has been completely unaffected by these revelations. Apparently their actions were different from those of the allegedly rare 'rotten apples', mentioned above. One recent commentator on Mendel's work has complained of the damage done by 'unnecessary' defences of the great man against accusations of fraud that have not in fact been made.[6] The implication is that careless commentary on science has associated him with the unquestionable guilt of others. Yet reasonable as that seems, what is the dividing line ? If we cannot specify how to make that distinction, then how can we say that major or real fraud in science is rare?

Here historical case-studies are of real value, as I hope to make clear. My discussion will be based on a distinction between what I call 'soft' and 'hard' fraud. The allusion to our conventional categories for the classification of pornography and non-medicinal drugs is deliberate. One purpose in so doing is to try to orient the discussion towards practices that are widespread and culturally recognized, if not necessarily sanctioned, and away from individual pathology.

We tend to use the term 'soft' to connote substances or representations that are in common circulation, but which are not thought particularly dangerous or problematic. We use the term 'hard' to denote a different class of substances or representations, to which much stricter rules should apply. Obviously the distinction here is fuzzy and shifting; to some people it is also irrelevant, or even dangerous, precisely because it carries the implication that there is nothing problematic about 'soft' pornography or 'soft' drugs.

Nonetheless this contentious classification is in very common usage, and much important discussion in law, criminology, social policy, political theory, and moral philosophy draws upon it. For example there is the question of whether the use of certain soft drugs, like marijuana, is in itself likely to lead individuals to more dangerous materials like heroin. Or there is the question of which kinds of videos can legitimately be prohibited from sale, in the public interest, without infringing the rights of individuals to consume whatever cultural products they see fit. These categories are at work in our culture, whether we think them robust, adequate or misleading.

There are at least three general issues here: one is the question of whether people have to be protected from the possible effects of indulging their desires; the second is whether the likelihood of damage to others is sufficient reason for prohibition; and the third is how people's sensibilities should figure in social policy. How for example do we resolve the contradiction between the outrage of some and the liberty of others? Empirical evidence has some relevance to the first two of these, although the many different claims about cause and effect are notoriously hard to resolve. This makes the question of how individual needs arise, are promoted, and legitimized centrally important. This is as true of the need to be recognized as a successful scientist as it is of the rather different need for physical pleasure through the injection of narcotics.

The 'hard/soft' classification can be usefully applied to fraud in research for several reasons. Firstly questionnaires and anecdotal evidence reveal that minor deviations from strict methodological propriety – 'improving' one's data, plagiarism of marginal ideas, poor statistical analysis, failure to cite important sources, and the prejudicial misrepresentation of others' work – are very common, whilst the classic cases of scientific fraud – invention of an entire set of data, possibly in many papers, deliberate creation of substitutes for evidence, wholesale plagiarism of others' contributions – are fairly rare.[7] On the face of it there appear to be two different kinds of activity that require labels.

I shall use the term 'soft fraud' to refer to those actions which have the effect of improving the evidence available in support of a scientific claim, or improving a particular individual's claim to authorship or originality, the assumption being that some grounds for the claim exist, for example that some empirical data has already been collected by the individual concerned. I shall use the term 'hard fraud' to denote those cases where there is nothing that would ordinarily count as evidence for such a claim and where it has been invented. The meaning of these terms is thus based upon and relative to the conventional standards of scientific evidence in particular fields at any given time.

Secondly both kinds of action are morally significant, albeit to a different extent. Soft fraud in science is generally said to be reprehensible, even though it is known to go on all the while. Thus whether many scientists think it really needs to be controlled is unclear, although the function of the refereeing and peer review system is precisely to maintain standards in this way. Hard fraud, on the other hand, is regarded as a deeply troubling phenomenon that must be controlled by the profession, because of the damage to the public image of science. It is precisely for this reason that senior scientists have felt it important to say that hard fraud in science is rare.

However, there is evidence that scientists are reluctant to do more to control minor misdemeanours, for example by spending more time on reviewing others' material before publication or repeating experiments, because it is not in their interests individually to do so.[8] It costs time and does not pay status. This maintains the conditions within which those resolved to take major short cuts to the top feel it is possible that they will survive. Equally journal editors and granting agencies are understandably reluctant to operate policies of systematic distrust of those submitting articles or grant applications to them. On the other hand, there are measures that would help. These include multi-centre investigations, where different groups have to assess their collaborators' results, tougher policies on the retention and provision of primary (lab bench) data, internal audit procedures, and random external review. The argument that soft misdemeanours are ineradicable or harmless – compare the arguments about soft pornography – should not be persuasive.

Thirdly it is clear that cheating in science, on whatever scale, is partly a psychological phenomenon, as individuals address felt needs to experience the pleasure of discovery and its reward. This has its roots in the earliest stages of science education, where we learn that certain kinds of cheating are acceptable on some occasions at least, to obtain the results that we were supposed to get, if necessary by retro-calculation. Arguably the pressures to do so, in school and university, are at their strongest, and the sanctions are weakest. Thus the process of becoming a fully socialized research scientist is one of learning to *stop* cheating in specific ways, despite one's prior educational experience.

Fourthly it is interesting to ponder the fact that both science and pornography are essentially about the representation of facets of reality normally hidden from view. In the case of science theoretical speculation and experimental analysis are intended to produce representations of processes, mechanisms, and structures not accessible through direct experience. Thus we use research equipment such as microscopes or telescopes to gain access to new domains or we use theories to codify our ideas about the regularity of natural processes, conceptualized within general simplifying schemes. We have first to represent nature in specific ways, in order to have scientific knowledge of it. Pornography conventionally offers us images of what are usually private acts or simulations of normally inhibited desires. However in both cases there are critical arguments that the representations in common use simplify, distort or otherwise misrepresent the phenomena, in order to gain mastery over them. Moreover Brian Easlea has drawn our attention to the use of terms in seventeenth century discourse about scientific investigation that evoke the idea of violent sexual possession.[9]

It must also be said that the analogy with drugs and pornography also breaks down in certain places. For example both these goods are supplied by an industry with scant respect for the law, or for the public good. Any sensible discussion of either has to take account of this fact. With research fraud on the other hand, we are not talking about something obtained from an external supplier, but about the internalized standards of autonomous individuals. Furthermore in the case of pornography, an important set of arguments against it concern its indirect effects on women through the consolidation of dangerous images of male and female sexuality. The effects of pornography on social conventions and practices work themselves out in very different ways, even though one might argue that a tolerance of soft fraud in research, by protecting science from more critical public scrutiny, preserves male power.[10]

The Case of Robert Millikan

Let us now consider a historical example of minor deviation in research practice. I have chosen this example for several reasons. The experiment is well known and often used in physics teaching at school or university. It is one in which the 'answer' is well known and it is therefore quite possible to cheat. Some students certainly do, with some justification. That in itself is an interesting issue. Also it is a case where the experimenter's mildly deceptive actions were actually more serious in context than they appear now, particularly as we 'know the answer', or think we do. This also poses very interesting questions about historical interpretation and reconstructing the original context of justification.

However my purpose is *not* to press the charge of professional malpractice against the man concerned, but instead to suggest that his pragmatic 'improvement' of his data to secure an intellectual position is actually very common in science. Our educational practices need to take account of this fact. Students need to be taught that science is about making a case for particular points of view, in circumstances where the data alone is insufficient to decide the issue, where the meaning of that data depends on prior theoretical commitments, and where its usage and evaluation is conditioned by a whole range of factors, not just the idealized rules of scientific method. Soft fraud is just a particular form of the argumentation characteristic of science, in which the rhetorical moves are, strictly speaking, illicit. However the argumentative moves in soft fraud are wrong because they are undeclared and lack the kind of justification they purport to have, not because they are rhetorical.

My historical example concerns the work of the American physicist, Robert Millikan, on the measurement of the charge on the electron. My remarks are entirely based on the work of Gerald Holton and Allan Franklin.[11] In 1907, Millikan, then a researcher at the University of Chicago, with several textbooks to his credit, but no significant research achievements, took up the investigations that laid the basis of his power and fame in science and brought him the award of the Nobel prize in 1923. Using a cloud chamber developed by C.T.R. Wilson, and already used by him in an attempt to measure the electronic charge, Millikan published results broadly consistent with the values of this fundamentally important physical quantity obtained by others, Ernest Rutherford amongst them. In so doing he placed himself in a position where as an experimentalist considerable professional rewards could be won.

The basic procedure was to observe the motion of a charged cloud of water droplets moving in the electric field between the parallel plates of a condenser. On the assumption that each droplet was equally charged, the value of the smallest amount of charge that any droplet could acquire – the charge on the electron – could be calculated from the rate of motion of the cloud. However it was theoretically possible that electric charge was not quantized, but could range over a probability distribution of infinite extent. Experiments like this one would thus measure an average, not an elementary value. Only if one was prepared to accept the postulated existence of microphysical particles like electrons – an assumption that seemed gratuitous and unnecessary to some physicists at the time – did the idea of particulate charge make sense. To Millikan however this did seem eminently reasonable. A further problem with the procedure in this form was the likelihood that the droplets in the cloud were becoming smaller by evaporation as the experiment progressed, changing the forces acting upon them.

Thus in the spring and summer of 1909 Millikan set out to explore this problem in detail, and in so doing discovered a radically new method. When he increased the voltage across the condenser he found that all but a very few droplets had been whisked away. However instead he was able to observe individual droplets moving up or down, or at rest. He had in effect an incredibly sensitive balance, with which he could measure the electrical forces created by the acquisition of units of charge by single droplets. More exactly, if the measurements with this apparatus showed that the forces acting on the droplets increased by discrete and equal increments, then it was extremely likely that such increments were a direct measure of the unitary charge on the electron.

His first results were described to a very distinguished audience at the meeting of the British Association for the Advancement of Science in Winnipeg in the autumn of 1909 and published in December of that year and in February 1910. In his papers Millikan reviewed the values obtained by other physicists, for purposes of comparison, excluding those he believed unreliable. One man whose work was thus set aside was the Austrian physicist, Felix Ehrenhaft, a young but highly regarded investigator, who had the distinction of being the first to publish measurements based on the motion of individual particles. Ehrenhaft's calculations depended on some debatable assumptions about his apparatus, although his values were in fact closer to Millikan's than some Millikan rejected. Rather remarkably Millikan's 1910 paper in the *Philosophical Magazine* contains the frank account of how he graded some of his results, apparently quite subjectiveley, for their reliability, and of the fact that he had rejected others, either because he had reason to suppose his equipment was malfunctioning or simply because they were inconsistent with earlier, credible results.

Ehrenhaft was apparently irritated by this slight on his work, malicious or otherwise, and began, with his students, to publish a whole series of papers containing more data. In the spring of 1910 he made the remarkable claim that his observations indicated that the elementary value of the electronic charge had to be significantly lower than previous work had indicated. Moreover he postulated the existence of a 'sub-electron' to account for this fact. Furthermore he drew support for this idea from Millikan's own published data, which Ehrenhaft recalculated on a slightly different basis. The effect was to produce a whole spread of values, where Millikan had obtained a tight cluster around a mean close to earlier results. For the next 13 years the two men engaged, first continually and then sporadically, in a controversy about the properties of electrons.

Millikan for his part had new data to report by mid-1910, obtained using a second modification of his basic method. On Holton's account what is striking about this publication is Millikan's confidence in deriving a mean value from a sub-set of observations derived from many runs of his equipment. Whilst Millikan does not conceal this fact, it is not made clear why some of them were excluded from the final computation. A further major paper in 1913 offered yet more data using more sophisticated equipment. For this work Millikan's original notebooks exist. It is these that have been used by Holton and Franklin to check the evidential basis of Millikan's claims and to explore why he felt able to exclude some results, whilst Ehrenhaft felt able to use similar – or, as we have seen, the same data – to argue in a completely different direction. Eventually the Viennese scientist chose to abandon the atomistic hypothesis of the existence of elementary particles altogether.

The re-examination of Millikan's results indicates that, despite his very definite statement in his paper of 1913 that he had used every result obtained over a long period of time, this was not the case. Whilst he did indeed use some of the observations made on the set of 58 droplets that he claims to have observed, it is not the case, as Holton demonstrates, that he used all of the observations of all of the droplets. The effect is to make more credible his value for the charge on the electron and to make less plausible the alternative possibility of fractional charges advocated by Ehrenhaft. Millikan believed it justified, and indeed necessary, to 'suspend disbelief', as Holton puts it, about the hypothesis that ostensibly he is attempting to test. In some cases his notebooks indicate why he thought it reasonable to exclude an aberrant observation: in other cases there are no such clues.

Franklin's re-examination of Millikan's notebook calculations leads him to slightly different conclusions. Whereas Holton only followed Millikan's data analysis and noted its mixed and censored results, Franklin has actually recalculated sets of final values for the charge on the electron from Millikan's raw data, excluding those cases where Millikan gives reasons for abandoning a measurement or where it is plausible to assume that there was a good reason because the kind of deviation from the expected value resembles that 'explained' away in another case. This is a generous but plausible approach to Millikan's data analysis, bolstered by the general argument that he had more data than he needed with which to compute the published value. He could indeed be selective. The question is how did he go about it and what effect did it have.

Franklin demonstrates that in 30 out of the 58 experimental runs on which the published results are based, some 'cosmetic surgery' on the data has gone on. Whilst it is possible to argue that the overall effect on the mean value of the combined experimental error values is slight, it is clear nonetheless that Millikan misrepresented the basis on which his data analysis had been performed and that he repeatedly chose computational procedures that gave results consistent with his expectations, whilst glossing over this fact.

At the same time it is as well to remember that these results were produced and analysed after almost six years of experience with the apparatus and attempts to reduce its sources of error. Moreover his results from February 1912 show less and less variation. He had good reason to assume that he could identify some discrepant results. Ehrenhaft on the other hand seems to have refused to consider that any of his own results could be considered discrepant. Anything registered by his apparatus had to be included, which is somewhat bizarre. Furthermore there is no indication whatever that Millikan modified his raw data and fabricated measurements that were never actually taken. But one cannot avoid the conclusion that he treated the results at his disposal in ways that he knew would render them slightly less vulnerable to potential critics and that he did so covertly. No doubt many thousands of scientists have done the same, but historians only bother to go back to the notebooks of the most distinguished. In Millikan's case one could not argue that he only achieved that distinction by improving his evidence. After two unremarkable decades in science he picked the right problem and set to work. Having no time for alternative explanations of his results, he exploited his opportunity with vigour and determination.

Conclusions

Millikan's actions are clearly different from those of John Darsee, Mark Spector or Michael Briggs in recent years. Some might object to my use of the term fraud in this connection, as a form of *lèse-majesté*. I acknowledge the disparity between the two kinds of activity and the larger one between the two sets of consequences: the consolidation of a major theoretical question and the measurement of an important physical constant in Millikan's case, and confusion, expense, and distress in the case of the others and many like them. Nonetheless I believe that Millikan's data analysis was soft fraud as I defined it above. Its minimal effects on the course of research were contingent upon the outcome that is known to us only now, but which at the time was far from clear. Ehrenhaft now seems a misguided controversialist, but to many of his contemporaries he was a clever and imaginative theoretician. By contrast Millikan appears to us now as a sober pragmatist. But the fact that Ehrenhaft's use of his data seems to be rather perverse does not exonerate Millikan for the way he played his hand in the controversy.

Interestingly Holton suggests that Ehrenhaft's final move to the Machist anti-atomist camp, the final step in his epistemological volte-face over the existence of micro-physical entities was opportunistic, part of an unsuccessful attempt to gain a chair. Whether Millikan was aware of comparable opportunities open to him I cannot say. Nonetheless in many cases experimental or theoretical success is traded for professional advancement, in terms of resources, status, and power. *En route* it is frequently tempting to play one's hunches, and sometimes to go a step further by 'improving' one's data.

Whilst soft fraud has relatively low benefits, it is also often a low risk tactic; hard fraud on the other hand may have significant advantages, but it is a very high risk option that spells the end of one's career if discovered. After all Millikan's very minor deviations were only discovered 20 years after his death and 60 years after the event. No reviewer would have been able to spot the problems, but perhaps his students knew? There again perhaps they were encouraged to think that this is just how one does research? After all that is a reasonable inference from both the evidence of many research papers today, that on careful review contain numerous minor errors and incomplete arguments, and from the teaching of science, where there are countless informal cues to adjust one's observations to the textbook account, and even in the limit to do so by retro-calculation from the accepted result.

Holton's and Franklin's papers are important contributions to the history of science. Inevitably they focus more on Millikan than on his erstwhile rival. They also pose new questions. For example I have stated at several points that Millikan knew what he was doing. Presumably this is the case: one can scarcely shift between computational procedures in one's data analysis without being aware of it. However his actual intentions are obscure, particularly, as Franklin argues, the effects of his data improvement are minimal. If he knew this why did he do it? But then if he did not know their effects why did he do it? Also assuming that we could document an informed awareness by Millikan of the arithmetic effects of his actions, can we then assume that he knew that his raw data would never be scrutinized? Can we also reconstruct the conventions of the time and deduce from them what Millikan might have thought the rhetorical effects of less dispersed data would be? What kinds of errors were thought tolerable? What kind of consistency would have been his goal?

This picture of 'tolerable' or 'understandable' soft fraud prompts a series of other questions about the genesis of hard fraud. Is it the case for example that those eventually found to have committed such acts do so after years of soft fraud? Or do they perhaps have abnormally high standards of proof, which they are unable to meet in a legitimate way? Have they internalized unrealistic ideals of performance, and, lacking faith in their own abilities, opted instead for a kind of instant gratification? Can one trace patterns of influence across generations of investigators, with people in effect being taught to commit soft fraud? At what stage in research training are such cues transmitted?

Selected case studies in the history of science offer a way of opening up rich and complex questions about contemporary scientific practice and education. They require us to ask what is learnt in science education and how things that we may not intend to teach can be brought to the surface, acknowledged, and dealt with. But the exemplary use of history is only one way of approaching these issues. Another, in many ways more appealing to students and teachers, is to use a simulation game, such as the one I have developed as part of my teaching in the history and philosophy of science in Manchester.

The Fraud Game is a board game that simulates aspects of a scientific research career. It is intended for 8–12 players, although it is quite possible to play with more or less. Each player moves or acts in turn on the throw of a dice, around a rectangular board, on which the squares represent possible laboratory sites. The object of the game is to accumulate status points through imagined scientific publication. Players may only publish their own papers on their turn. To publish papers requires the expenditure of resources, which are simply set aside in an envelope. These units of resource are replenished from the bank after each cycle of turns. The quality of any paper depends in a non-linear fashion on the quantity of resources devoted to its production. Thus one trivial paper costs one unit of resource; one good paper costs 10 units of resource; one outstanding paper costs 100 units of resource.

Whilst the quantity of resources needed to produce a paper of a given standard is known to all players, the actual quantity utilized is a private fact known only to the player concerned. But of course players may cheat by using less resources in any publication than they claim to have done. Unless they make this very obvious, by cheating too early in the game or by publishing very good papers much too often, this will probably remain undiscovered. Yet the rules allow for challenges, when the truth will come to light. There are two kinds of fraud, soft and hard, defined in terms of the quantity of missing resources. But to make a challenge itself requires units of resource from the challenger, and the rewards from a challenge that brings fraud to light are minimal. In other words it only makes sense to challenge if it is certain that one player is improving his or her chances of winning through serious fraud.

On the other hand if you are discovered as a principal author to have committed a minor fraud you lose up to five units of resource in the next cycle only. This is tolerable. If you are discovered to have committed a major fraud, your status falls to one. This is likely to be catastrophic. Moreover if you have a lab, you lose it and must move on. Publication costs are then doubled for the rest of the game for you, and any collaborators, if you can get them. This applies on each occasion on which you are discovered to have committed fraud.

Under certain circumstances players face dilemmas which have resource and status implications. As in other games they select a card from a pile and read out the choices open to them. They take a decision and the banker reads out the consequences from a directory. The dilemmas are all based on known cases of scientific fraud, where colleagues or the individuals concerned had to face difficult choices. These dilemmas are educative in themselves and can serve as the basis for further discussion.

Finally status changes upon publication, according to predetermined rules. So too does the amount of resources allocated to each player, which depends upon his or her status achieved by that stage in the game. The formulae here are somewhat complex, and have been determined by trial and error over many games, to get the right gearing between status increase and allocation of resources. With the wrong ratios people play for ever without being able to publish, or quickly realize that simple ploys will guarantee ludicrous amounts of status in a few moves. The rules also allow for collaboration and for identifying fraudsters within such collaborations.

The rules take about 15 minutes to explain, and the average game lasts about one hour and a half, although it can go on for longer if the players wish. It may thus make sense to spread the game over two sessions, if only limited time is available, devoting only the second to serious play without prior explanation of the rules. The rules are actually quite simple to grasp but slightly harder to use tactically, just as with many other games. It requires no knowledge of science other than that some scientists do experiments and that they publish papers.

What emerges is that hard fraud is a very high risk tactic, at least if one has accumulated some considerable status, whilst occasional soft fraud may be low risk and make tactical and strategic sense. Responding to indications that hard fraud is occurring is in most players' interests, although not all will feel able to pursue these interests by making a challenge. As one might expect players with more resources and more status are freer to act and have more to lose by tolerating cheating. Nonetheless accepting that soft fraud is probably going on may also make sense, since the gains in uncovering it and the advantages in pursuing it are minimal. This of course encourages the desperate to contemplate hard fraud more seriously.

At present only a few copies of the material needed to play the game exist. This includes a playing board, a set of rules, a score sheet, and a set of dilemma cards. If there is significant interest in *The Fraud Game*, then more copies of a better quality could be produced for a small fee. Inquiries should be addressed to the author at the Department of Science and Technology Policy, University of Manchester, Manchester M13 9PL.

NOTES

1 Medawar (1969).
2 A fascinating example of the historical character of basic scientific conventions is Shapin's analysis of the way in which seventeenth century investigators sought to demonstrate the credibility of the observational reports: see Shapin (1984).
3 Yoxen (1985).
4 See for example Broad and Wade (1982).
5 For a recent review of some of the more analytic literature see Chubin (1985).
6 Edwards (1987).
7 Chubin states that the National Institutes of Health in the United States has claimed that between 1980 and 1982 45 cases of alleged misconduct came to their attention, a figure which is to be viewed against that of 20 000 NIH grants and contracts active in any year: see Chubin (1985).
8 Chubin (1985); see also Anon. (1987).
9 Easlea (1981).
10 One could argue for example that this occurred in the Briggs case, where largely, or perhaps wholly, fictitious data was used to market new contraceptive products and was both presented and accepted as 'authoritative' and reliable evidence.
11 Holton (1978), Franklin (1981).

Sources and Resources

Textbooks

David Knight

More than twenty-five years ago, as a chemistry graduate just turning towards the history of science, I went to a conference in Oxford at which Thomas Kuhn talked about dogma in science.[1] His *Structure of Scientific Revolutions* was written but not yet published, and in this lecture he spoke about scientific textbooks. He said that while in the humanities one might find a teacher sad or furious that a class had used the 'wrong' books and got a hopeless picture of how to go about criticizing a text or whatever; in the sciences textbooks differed in their up-to-dateness and in their clarity, but not seriously in their content. He saw science teaching as essentially dogmatic. Historians would be expected to wrestle with conflicting interpretations, but students of the sciences would not. Theirs was a training, not unlike the drilling of recruits who after a few weeks on the parade-ground will do what the sergeant orders without question on the battlefield. In science there are many ways of going wrong, and it is very important that the student should not be tempted towards one of these broad paths leading to destruction, but should stay on the straight and narrow way which eminent predecessors had found to bring success.

For Kuhn it was therefore not an accident that science teaching was dogmatic; it was the inculcating of a 'paradigm'. Like other parts of Kuhn's writings, we may well feel that this is a caricature; but caricatures are visions of truth, distorted. There is enough truth in his picture to remind us that textbooks in the sciences are not personal documents, and that the 'heuristic' or 'discovery' method of teaching must be to some extent a matter of play-acting: the pupil must discover oxygen by 3.45pm, because next week we get on to chlorine.

The science in textbooks is thus established science; the sort of thing which interested Hegel,[2] concerned about levels and transitions, although most philosophers of science have been more interested in discovery and justification. Among textbooks we may find some great books of science, but on the whole they are not original, and are of interest to the historian on this account. They show the norm, and they illuminate the teaching of science in the past. To be a textbook a book ought to be closely tied to a course, and very likely to an examination. Euclid's *Geometry* is probably the most successful of textbooks, though probably not written as such, and some great books may become a kind of textbook; thus at Cambridge in the early nineteenth century there was an exam paper just bearing the title 'Newton', based upon the early parts of his *Principia*. No doubt most of the undergraduates worked it up mostly from more recent textbooks. To be up-to-date textbooks need frequent revision; the most successful have been through numerous editions, which it may be interesting to compare. If we look at a recent bio-bibliography of eighteenth-century mathematical works,[3] we find textbooks surprisingly prominent, and selling in what must have been large numbers.

Lavoisier's *Elements of Chemistry* (1790) is almost like a textbook, but he was not a teacher in an institution and it was not aimed at an existing course; it might be used, as it was by Davy, for self-education. This is a work of rhetoric; the reader comes to see chemistry willy-nilly through Lavoisier's eyes as he reads the text and tries the experiments. It is therefore a classic,[4] and very original; but not quite a textbook. There were real textbooks in chemistry, especially associated with medical education. Boerhaave at Leyden found his lectures being published in a pirated edition, and duly wrote them up into a textbook;[5] both the pirated and the authorized versions were translated into English, as the Leyden model was adopted at Edinburgh and at Glasgow. Thomas Thomson's textbook,[6] written for medical students in Scotland, was striking because Dalton's atomic theory was first published in it – a curious inversion, preceding Dalton's own presentation in a monograph and in papers.

In the early nineteenth century, with the Apothecaries Act of 1815, formal courses in chemistry became essential for medical students in London, and textbooks were duly written to accompany the lectures. Davy at the Royal Institution had written up his very successful chemical lectures as *Elements of Chemical Philosophy* in 1812, but this was not a textbook, being selective and speculative. His successor W.T. Brande delivered courses of lectures to medical students, and wrote a textbook, while both Turner at University College and Daniell at King's

College also produced textbooks for their courses.[7] What is interesting is how different these latter two are: Turner was an analyst and his book is in the Scottish tradition where analysis was central, while Daniell was a close associate of Faraday, and in his book what we would call physical chemistry is all-important. Here we have the modern situation in which students, many of whom will not have been especially interested in chemistry, have to get through a syllabus and pass an examination if they are to get a professional qualification. In later chemistry textbooks we can see how the 'pure' parts of the science came to prevail over the 'applied'.[8]

Because there were not in the early nineteenth century school syllabuses in chemistry, there were no textbooks in the strict sense, but there were books written for children. One was Jane Marcet's *Conversations in Chemistry*; and another was Samuel Parkes' *Chemical Catechism*,[9] both first published in 1806. Jane Marcet turned Faraday towards chemistry, while another of her books got the young J.S. Mill interested in political economy, which is not a bad record. The books reproduce tutorials, with Emily, Caroline, and Mrs B., their tutor; they are not girls' books in any way beyond this, and their success should make us hesitate in snap judgements about women's education. Parkes' book was also originally written for a girl, his daughter, but his address was the Haggerstone Chemical Works. His book is a compendium of useful knowledge, accompanied by numerous reflections on Providence – he was a zealous Unitarian. Readers of Mrs Marcet learned chemistry as exciting science, while Parkes' readership was expected to be young men needing information.

Lavoisier's book marked a revolution in chemistry, but during the nineteenth century, perhaps early[10] and certainly by the middle with conservation of energy, the new science of physics came into being. This was certainly a social construct; mechanics, optics, electricity, magnetism, and other sciences which had been differently grouped were brought together. The new science needed a textbook which would start from conservation of energy, and William Thomson (later Lord Kelvin) and P.G. Tait, very eminent physicists at Glasgow and at Edinburgh, set out to write one.[11] The result, affectionately called 'T + T', came to a halt when the two authors could not agree on the value in physics of the mathematical functions called quaternions. Despite these problems, we see in this book an advanced textbook to accompany the specialization which was a feature of the nineteenth century and since, accompanied by a broader vision which brought separate sciences together.

At about the same time Darwin achieved a similar revolution in the life sciences, and while the *Origin of Species* (1859) is very readable and, deceptively easy, it is not a textbook. Darwin's 'bulldog' T.H. Huxley taught courses at South Kensington, and in 1879 produced a textbook, *The Crayfish*.[12] If we seek for an overt evolutionary perspective throughout we shall be surprised, for only the last 10% of the book is concerned with the possible origin and development of crayfish and other crustacea. Huxley's course, as opposed to his public lecturing, was not strongly Darwinian. What makes the textbook a classic is Huxley's use of the crayfish as a 'type'; instead of learning a little about all sorts of creatures, his students studied the crayfish in exhaustive detail. There is even a chapter, very short, on the crayfish mind.

Textbooks are themselves physical objects, and the change in their appearance over the last twenty-five years or so is very striking. A previous period of very rapid change was around 1830, when steam presses, cheap paper, publishers' cased bindings, and illustration by wood-engravings and lithographs instead of copper-plates, transformed the book trade. Cheap books went with the March of Mind; Faraday got out of craft bookbinding at about the right time; and we can from old textbooks learn some social history – especially if they have some publishers' advertisements bound in at the back. Parkes used big type for what had to be learned and small type for the explanations which one might not need, and in his and later works we see attempts to express chemical mechanisms using ingenious typography. Illustrations of apparatus become especially interesting when, as with those in s'Gravesande's Newtonian textbook from Leyden,[13] they can be compared with actual pieces of apparatus preserved in museums; in this case with those at Leyden and in the collection at the Science Museum in London which was put together for the young George III.

Some of the great textbooks contain a good deal of history; Brande's, for example, and Thomson's, although he wrote a separate *History of Chemistry* (1830). British authors tended to use history in urging that Boyle rather than Lavoisier was the founder of scientific chemistry, but also to legitimize their view of the mainstream of the science. The great example of the use of history was Charles Lyell, whose *Elements of Geology* (1830–3) was almost a textbook; he taught at King's College London for a time. There he argued that his predecessors had been prodigal of violence because they had been parsimonious of time. Such cases can remind us how selective a business history is, and how there is no one history; the judgements and prejudgements of historians cannot be avoided.

All this may indicate that old textbooks on whatever level are an important resource for the historian of science, and that there is still much virgin territory waiting exploration. In the history of science we do not have textbooks in the same way as the scientists do, for the reasons Kuhn indicated. There are courses in a number of universities, many of which will survive the UGC exercise in 'rationalization' published in the summer of 1987. They need reading lists, but one wants students to come up against conflicting interpretations. Books which present a new synthesis are therefore really desirable, to come between the student and the monograph or the primary source. These are not exactly textbooks, and indeed one's departmental 'profile' would not be improved by the writing of textbooks because of the belief that there is no element of 'research' about them. The best general introductions are based on both primary and recent secondary materials, like papers in journals, and are not just popularisations.[14] In this sense, we should be glad that there are few textbooks in the history of science.

There are on the other hand many books, and some articles (and certainly book reviews and essay reviews, although we are short of Macaulays) in journals which are perfectly accessible to teachers and one might hope interesting to them; although they are in no sense an attempt to present the whole truth prepackaged. Students in higher education at least can be expected to wrestle with primary sources; indeed in the history of science this is analogous to laboratory work in the sciences. Schoolchildren are not probably in this position; illustrations and brief extracts can bring things to life, but great chunks unmediated do not seem appropriate. So some kind of textbooks are needed; and it might be that textbooks of history of science for children would have a place, like those discussed by Joe Scott in this book. Here those teaching in universities will have to repress their feeling that they would rather encounter students with open minds. Schools can be magnificent agencies for inoculating the young against various disciplines, and those of us teaching 'non-school subjects' count ourselves in some way as lucky, but it is really a good thing that large numbers of 14 or 15-year-olds should have the chance to learn some history of science, although of course it cannot be approached with the sophistication appropriate to those of 18 or 19.

Here the difficulty is that over the last twenty-five years or so the writing of the history of science has changed direction. Anyone of restless and versatile intellect will find that his or her own works move with what may be just fashion but also represents genuine changes of interest. The old 'internal' history concerned to establish who did what has given way to a concern with 'external' factors, and this makes science look much more of a fully human activity. Historians have begun to make much more use of the archives of scientific institutions, partly because they are there, partly because of a belief that manuscripts are 'better' than printed materials, and partly because illuminating careers and social mobility can seem more exciting than reconstituting old theories and explanations. The historian is bringing to life a valley of dry bones, and he has to see science as a practical and a social activity as well as an intellectual one. Science also gets a worse press than it used to do, perhaps; certainly it has become easier to see scientists as careerists.[15]

Historians have produced exciting studies of the Royal Society and of the British Association in their early days, for example; have looked at the gentlemanly ethos of pre-Victorian geology; and have written up the steam-intellect societies concerned with the diffusion of science amongst artisans.[16] Indeed one can perhaps see historians of science as engaged in a series of flirtations, with scientists, with philosophers, with historians, and with sociologists.[17] In each affair, different aspects of that complex activity we call science seemed important. The old question 'What did X do?', best answered in a simple way from a biographical dictionary,[18] became much harder to answer in a sophisticated way, just as the older one 'Who discovered X?' had. The placing of people in contexts is central to the historian's activity, and writing biographies of scientists is thus a major activity among historians[19].

For someone writing a science textbook, with a course and a syllabus in mind, and with the hope of sugaring a pill or of demonstrating that science is done by fairly ordinary folk whose motives and thoughts can be followed, there is a problem of what history to put in. Very likely the anecdote, or the occasional case-study, is the best answer; the attempt to cover too much ground can produce something very dry. One might expect historians to be more interested in what it was like to be a student of science a hundred years ago, and chemists to be more interested in how the first synthetic dye was prepared; but our experience in Durham with students taking elementary and advanced courses in history of science, from different backgrounds and taking different degrees, indicates for what it is worth that this need not be so. The history of science is rich in alternative perspectives, instructive failures, changing career-patterns, and different kinds of utility, and it would be very good if these things could be got across in science textbooks, as well as in history ones, where the history of ideas is at present neglected. But this will have to be done by those familiar with courses; those outside may produce readable works, but cannot really prepare textbooks.

NOTES

1 T.S. Kuhn in Crombie (1963). See also King-Hele (1985), p. 260.
2 Horstmann and Petry (1986).
3 Wallis and Wallis (1986), Becher (1986).
4 Gjertsen (1984). His rather curious list of twelve works does not include Lavoisier's.
5 The authorized English version was Boerhaave (1735). See also Underwood (1977).
6 Thomson (1802); by 1831 it had reached its seventh edition. Donovan (1975).
7 Brande (1830), Turner (1831), Daniell (1839).
8 Bud and Roberts (1984).
9 Knight (1986); Marcet's eleventh edition came out in 1828.
10 Home (1983).
11 Thomson and Tait (1879), Harman (1982), Cantor (1983).
12 Huxley (1880).
13 s'Gravesande (1747), Ellenius (1985), G. L'E. Turner (1983). Turner's persidential address on scientific toys is forthcoming in *BJHS*.
14 Accessible works for the seventeenth century include Cohen (1986), Hall (1983), Debus (1978), and Hunter (1981); for later see Russell (1983) and Knight (1986). All these have guides to further reading, and are available in paperback.
15 Kohn (1986), Franks (1981).
16 Morrell and Thackray (1981), Hunter (1982), Rudwick (1985), Inkster (1985).
17 Knight (1984, 1986).
18 *The dictionary of scientific biography* (Gillispie 1970–85) is a multi-volume compilation and a standard work, though weaker on natural historians than one might have hoped. See also Williams (1969) and Wintle (1962). *The dictionary of the history of science* (Bynum, Browne, and Porter 1983) is conceptual rather than biographical.
19 Examples are Brock (1985), Crosland (1978), Gillispie (1983), Gooding and James (1985), Jordanova (1984), and Russell (1986).

Journals and Societies

Paul Weindling

The past twenty years have seen an explosion of interest in the history of science, medicine, and technology. This interest is due to a greater public questioning of the ethics, costs, and implications of scientific advance. Movements as diverse as the US civil rights campaign for black equality, feminism, the anti-vivisection campaign – that itself has a fascinating history – and the demand for attention to ecology have all generated the demand for greater public accountability and understanding of the sciences. International scientific rivalry has also played its part, as when the launch of the first Sputnik caused consternation that the USA might be losing its scientific lead. Such political and social developments have meant that the history of science, medicine, and technology have gained in status as academic studies in their own right, and they have attracted academics with a variety of backgrounds in science, history, and the social sciences making them remarkable areas of inter-disciplinary endeavour. They have developed from amateur leisure activities of scientists and doctors to becoming the sphere for full-time professional historians; they have considerably broadened their appeal to academics in other disciplines as history, anthropology or sociology. These developments have resulted in the abandoning of a positivistic and achievement-oriented history of the subjects, and greater awareness of science, medicine, and technology as part of cultural and economic history. We have a richer understanding of science with new dimensions revealed by institutional, biographical, and cultural studies. The history of medicine and technology are no longer assumed to be merely aspects of applied science, but are seen as involving distinct issues, such as relating to 'alternative' medicine and Third World technologies.

Interested school teachers took a pioneering role in challenging the antiquated standards during the 1960s. Instead of seeing science as the positivistic accumulation of objective facts and emphasizing great men and great discoveries, there was revolutionary change of approach to developing studies of science and medicine in their cultural and social contexts. A substantial proportion of those now occupying senior academic posts in the history of science began their careers as school teachers. Research on the British Association for the Advancement of Science or on the relations between science and the Puritan revolutionaries of the seventeenth century exemplifies the shift of attention to public perceptions and the social role of the sciences.[1] Journals such as the *School Science Review*, the *Journal of Chemical Education*, *The Journal of Biological Education* and *Physics Education* contain some original and much neglected articles from this period. For example, an often overlooked classic is Alan J. Bennett's 'Mendel's Laws' in the *School Science Review*.[2] Such journals have continued to publish the very occasional historical study. As much of the energies of the reformers of the history of science went into developing new perspectives, efforts were channelled into revising standards, and the centre of intellectual gravity of the subjects shifted to universities. As a consequence the special needs of schools have been relatively neglected. Certain topics preoccupying today's historians of science – like Foucault or the history of eugenics – cannot be directly transferred to a classroom context, even though some insights might sharpen the critical faculties of any teacher with regard to such issues as sexism, racism or the social accountability of sciences. There are, however, no organizations where historians of science, medicine, and technology can meet to consider the educational implications of their academic research. History of Science has no equivalent to the Historical Association. There is hardly anything in the way of either institutions or periodicals specifically for teachers or school children, although some of what is currently available might be developed.

Institutions and Societies

Most national scientific institutions and scientific disciplines have groupings and often publications concerned with historical topics. It has been of particular interest while conducting research to visit various scientific and medical institutions as these tell one much about the public image that their founders wished to project. The Royal Institution and the Natural History Museum not only hold valuable historical collections, but are themselves historical creations of immense interest. The Royal Society convenes the British National Committee for the History of Science (which includes the history of medicine), possesses an outstanding library and archival collections, and publishes *Notes and Records of the Royal Society of London*. The British Museum (Natural History) also publishes a useful historical series. Many specialized scientific societies like the Linnean, Geological or Royal Astronomical Societies have libraries rich in historical sources, and their librarians often have historical interests and are active in the British Society for the History of Science or the Society for the Bibliography of Natural History. Some formal groups for disciplinary histories have formed, such as the History of Earth Sciences Society (in 1981), or for astronomy with the publication since 1970 of *Journal for the History of Astronomy*. The oldest of these is the Society for the History of Alchemy and Early Chemistry, which has published *Ambix* since 1937, and now has extended its range to include modern chemistry. An example of a recent initiative is one for the history of biology based at the Institute for Biology (organiser Nicholas Russell).

While the endeavour of any discipline to maintain awareness of its history is to be welcomed, it is the case that it is those groupings most open to outsiders that maintain the liveliest and best historical discussions.

Many issues relevant to the social history of science, such as professionalization or the demarcation disputes between emergent sciences, are relevant to understanding the rapid rise of the history of science and medicine as academic disciplines. As the histories of science, medicine, and technology have struggled to find their identities, each formed a specialist society, and these societies have themselves involved histories of their own. The British Society for the History of Science (founded in 1947), the Newcomen Society (dating from c. 1920), and the Society for the Social History of Medicine (founded in 1970) pride themselves on their inter-disciplinary and multi-occupational character, and like to be thought of as out-going evangelizing organizations. SSHM and BSHS each has over 500 members.[3] These societies publish journals and monographs, and organise regular meetings. The United States' societies have also been pace-setting. The Society for the History of Technology (SHOT) publishes the innovative and broad-ranging

Technology and Culture. The History of Science Society's journal *ISIS* is to be recommended, as also is the Society's annual bibliography of the subject, the retrospective cumulated bibliographies, and its occasional guide to the history of science that is packed with addresses and useful listings. The American Society for the History of Medicine publishes *The Bulletin for the History of Medicine*, which maintains impressive scholarly standards.

The history of medicine in the UK has been especially vigorous owing to the varied interests that have found an outlet in SSHM and to the support from the Wellcome Trust. Henry Wellcome envisaged a Museum of Mankind, which was to present an anthropological approach to the history of health and healing.[4]

Wellcome's vast collections and library spawned several organizations. It is worthwhile to comment briefly on institutions funded wholly or substantially by the Wellcome Trust, not least because they are often misunderstood to be all part of a single mammoth institute. In fact the Wellcome Institute for the History of Medicine does contain the largest specialist library for the history of medicine, an academic unit and a contemporary medical archives centre[5]. There is the Wellcome Museum of the History of Medicine, which has been transferred to the Science Museum, and a new Wellcome Tropical Institute which also contains historical collections. The Wellcome Institute has links to a unit at University College London, and there are academic units at Cambridge (in the Department for the History and Philosophy of Science), at Oxford (in the History Faculty), at the Department of Science and Technology Policy in the University of Manchester, and at Glasgow. These varied developments have resulted in a vigorous and innovative subject which is able to maintain broader perspectives in a university context.

As the concerns of historians have been extended to cover very recent issues – and in the case of AIDS there is an on-going historical project – the collecting of archives of living scientists and policy-makers is increasingly important. In the wake of the Sputnik's challenge to US scientific preeminence, the first US conferences on the preservation of scientific manuscripts were held. This resulted in special resource centres, the first being at the American Institute of Physics. Concern to preserve the records of eminent scientists resulted in an initiative by the historian of atomic energy, Margaret Gowing, and the Oxford physicist, Nicholas Kurti, in 1973. They established a Contemporary

Scientific Archives Centre in Oxford (now at Bath) to locate and process papers of scientists, and the idea has caught on with numerous specialist centres, particularly in the USA. The Department of Science and Technology Policy at Manchester is opening a centre for the history of computing, and there is a Charles Babbage Institute in the USA. Major institutions have also established resource centres, a notable example being that of the Rockefeller Foundation, which shaped the development of science and medicine in the 1920s and 30s.

Such initiatives are primarily oriented to servicing academic research. There has been far less attention to general public information or to the needs of schools. As university departments of the history of science and medicine are suffering from the current malaise of the under-funding of the universities, they are not in their current state particularly suitable resources for acting as resource centres for school children tackling projects, unless these departments have some special source or facility. Although the research unit where I am located does not have the resources to advise school children for their projects on any scale, we have given guidance to local schools for locally based projects. The situation is not helped by poor initial guidance from teachers who suggest topics that are so global as to be vacuous. A typical inquiry is that a pupil request guidance for a project on 'the history of medicine'. To be genuinely helpful it is necessary to have a precise and specific problem or study area; preliminary guidance on how to formulate answerable questions and practicable projects is a prerequisite. On the whole I would recommend as information sources the national institutions which have traditionally had a function as libraries or museums for the general interested public. Thus for the history of science, the obvious first port of call is the Science Museum. The Royal Institution Centre for the History of Science and Technology has since its foundation in 1799 concentrated on the popularization of science.[6] For the History of Medicine either the Wellcome Galleries at the Science Museum or the excellent Library of the Wellcome Institute for the History of Medicine might be approached. Pupils should be advised that local scientific and medical institutions are a resource that ought to be tapped. Given the wide network of mechanics' institutes and literary and philosophical societies, combined with the recent spread of industrial archaeological and heritage centres, there exists immense potential for local projects that can be combined with visits, fieldwork, and first-hand contact with the artefacts and documents of the past.

Periodicals

General science journals have often welcomed contributions on historical topics as can be seen from *New Scientist* or *Nature*. The innovative historical journals *Past and Present* and *History Workshop Journal* have given space to controversial new research. Yet at the same time such journals are necessarily limited in the amount of coverage that they can give. *History Today* has Roy Porter on its editorial board; he has made a valiant attempt to include examples of new approaches to both the history of science and medicine in a journal that has a strong school readership. This presents the history of science as part of history. The *Oxford Review of Education* has published special issues on the social history of science and on the history of special education, and *History of Education* contains many excellent articles relevant to the history of science, medicine, and technology.

The oldest British specialised journal is *Annals of Science*, founded in 1936. Other general journals are the *British Journal for the History of Science*, and its elder sister, *ISIS*, which has a history stretching back to 1913. The *Bulletin of the Society for the Social History of Medicine* is soon to become a fully-fledged journal published by Oxford University Press. The Newcomen Society has published *Transactions* since 1920. Historians of technology would be advised to study the US journal *Technology and Culture* as well as developments in industrial archaeology. There are a number of general journals which have a special function or approach. Examples are the review journal *History of Science*, which is ideal for in-depth reviews or research trends; the thematic *Science in Culture*; the US periodical *Journal of the History of Ideas*; and the journal *Social Studies of Science*, which is much influenced by the social construction of science approach as developed in Edinburgh by Barnes, Edge, and MacKenzie.

There are numerous specialized journals relating to particular disciplines (for example the *Journal for the History of Biology* or *History and Philosophy of the Life Sciences*) or periods. Other journals carry the output of particular countries, as do certain Australian or Canadian journals. Unfortunately, it is sometimes the case that the more restricted the scope of the journal, the lower the quality of the contents. However, some editors manage to run specialist journals of extraordinarily high intellectual quality. There are many more journals than any historian could possibly keep up with, and as a consequence bibliographical bulletins are used. An inexpensive way of keeping up with new research in the history of medicine is to read *Current work in the History of Medicine*, published by the Wellcome Institute; the National Library of Medicine (at Bethesda Maryland) publishes an occasional volume of listings for the history of medicine taken from the comprehensive *Index Medicus*. For the history of science the annual bibliography published by ISIS, and the cumulated ISIS bibliographies are useful.

When I counted up the numbers of history of science and medicine periodicals in 1982, I found nearly two hundred journals, bulletins and, newsletters.[7] Since then numbers have increased, owing to further specialisation, source projects, and specialist centres (as for the history of computing), and the readiness of publishers to invest in journals. Although there are some signs of improving standards, my impression remains one of quantity rather than quality. Yet at the same time the growing number of publications at a time of financial retrenchment in all branches of education suggest that there is a groundswell of intereset in the history of science, technology, and medicine. I cannot help feeling that a strong case could be made for some type of special publication for the history of science and medicine designed for schools and the broader public. For the history of science, medicine, and technology offers a way of promoting understanding of long-term trends, failures, and rejected knowledge in a way that teaches the need for a critical and responsible attitude to knowledge. Moreover, history shows that scientific achievement is the result of criticism and concern with broader social and intellectual issues.

NOTES

1. Modern classics produced by sometime school teachers include Morrell and Thackray (1981) and Webster (1975).
2. Bennett (1964).
3. For membership of the BSHS contact The Executive Secretary, Wing-Commander G. Benett, 'Southside', 31 High Street, Stanford-in-the Vale, Faringdon, Oxon; for the Newcomen Society contact the Science Museum, and for SSHM contact Mary Fissell, Wellcome Institute for the History of Medicine, London NW1 2BP.
4. For a history of the Wellcome Trust see Hall and Bembridge (1986).
5. The Wellcome Institute for the History of Medicine is at 181 Euston Road, London NW1 2BP.
6. The Royal Institution Centre for the History of Science and Technology is at 21 Albemarle Street, London W1X 4BS.
7. Weindling (1981).

Local Sources and Museums

Stella Butler

History of science and technology is a fairly long-established discipline with a wealth of texts, both general and specialist. And yet the discipline has not found its way into the school curriculum, despite its potential to broaden students' understanding both of the nature of science itself and of historical developments generally. Lessons about scientific and technical concepts can be humanized by learning the motivations behind and applications of particular discoveries. Students are often able to confront issues involving policy decisions more easily through historical materials. Examining the way scientists have worked in the past can often allow consideration of method and an evaluation of the role of evidence in scientific theory. Other chapters in this volume outline the resources in books and other media available to teachers and lecturers who are developing learning materials in this field. Museums offer unparalleled facilities to anyone aiming to stimulate students' interest in the past. In this chapter I outline the various types of science and technology museums in this country and review the ways in which these institutions and their staffs communicate to their various audiences and, in particular, how museums can contribute to teaching history of science and technology in schools.

Science and technology museums are many and varied in this country and can best be thought of in several categories. The Science Museum in London, the Royal Museums of Scotland, the Greater Manchester Museum of Science and Industry (GMMSI), the Birmingham Science Museum, and the Newcastle Science Museum all contain collections relating to many aspects of science and technology. Exhibitions cover topics such as steam power, electricity generation, air transport, optical instruments, and the chemical industry. These museums are large institutions and often provide temporary exhibitions which are well worthwhile.

The Museum of the History of Science in Oxford and the Whipple Museum in Cambridge, both university museums, possess large collections of very fine scientific instruments such as microscopes, telescopes, surveying equipment, and drawing instruments, generally dating from the eighteenth and nineteenth centuries. Neither institution is particularly interested in modern science, and neither collects or displays large technical machinery or large modern instruments. However, both museums offer a glimpse of the craft of the instrument maker through their superb collections. The Whipple Museum of the History of Science in Cambridge mounts a special exhibition on a particular theme each year. Subjects have included 'Specroscopy', 'The Social History of the Microscope', and a special exhibition to celebrate the tercentenary of the publishing of Isaac Newton's *Principia*. A set of attractive wall posters was prepared in association with the latter display and sold to schools and educational institutions. These posters, which were sponsored by Shell UK, provide an unparalleled range of scholarship, explaining Newton's life and work in a straightforward and attractive style.

Over the past ten years interest in our industrial heritage has increased dramatically. Several museums and heritage centres have been developed on sites of historical interest which have become available because of the changes taking place within the economy. The first of these industrial museums to be developed was the Ironbridge Gorge Museum. In one of Britain's most idyllic settings visitors can explore aspects of the iron industry and other technologies which formed the basis of the local economy. In the North East, Beamish also allows visitors to roam over a large area and see buildings associated with local industry and local life. A nineteenth century street has been reconstructed, for example. In the North West, a number of textile museums have been developed recently demonstrating both the technologies associated with cotton and silk manufacture and the kinds of working conditions the machinery operatives had to endure. These institutions include Helmshore Museum, Styal Mill, Macclesfield Silk Museum and Trencherfield Mill in Wigan.

Science Centres are the newest type of science museum. Exhibits are not kept in glass cases but visitors are rather encouraged to 'have a go'. The historical development of science is not generally stressed in this kind of institution. However, teachers looking for museum resources useful to the history of science should keep science centres in mind.

Their aim is generally to encourage exploration of the physical world. Although in some cases exhibits are not given any context, historical or otherwise, a visit to such an exhibition may be useful to set the scene for a subsequent class discussion, or for students looking at a particular technology.

Why Museums?

The term 'museum' as a self-explanatory institutional label is becoming less useful as we approach the 1990s. Institutions clearly vary enormously according to the emphasis placed on particular activities. However, fundamentally museums are about collections of objects. Those objects may have been brought together for a wide variety of reasons – either to record a major achievement, or because they belonged to a famous person, or to establish a representative sample of a discipline or a subject. Whatever the reason for the collection, objects are, of course, about the people associated with them and the concepts and processes of which they are a product.

Museums exist therefore to make collections of objects available and accessible to the public. Most institutions now recognize that people need help to make sense of objects and therefore provide contextual information in various forms within their exhibitions. Traditional object displays which generally involve glass cased objects explained by text and illustrations in the form of wall panels are the most common medium through which museums make available their collections. Often these exhibitions aim to present some form of coherent storyline. For example, in GMMSI's 'Microscopes in Manchester' the microscope collection is displayed in cases for visitors to admire and a history of the microscope as background information is provided through wall-mounted text. This kind of display is typical of museums throughout the country.

Recently, displays in the form of theatre sets have become more popular. For example in GMMSI's 'Collected Cameras' we have tried to express various aspects of the history of photography by constructing tableaux. One shows a local Edwardian craftsman's workshop with a beautifully crafted wooden field camera. This is followed by a set incorporating a Kodak Brownie camera in the context of the seaside with bucket and spade, etc. and snaps of popular holiday camps, in an attempt to get across the idea of a mass market for inexpensive cameras.

On an even more elaborate scale, complete rooms are recreated. Sometimes, a period 'feel' is all that is aimed at. However, serious points can be made: in the Electricity Gallery at GMMSI, a series of room sets demonstrate the state of domestic electrical technology in 1936, 1956, and the present day. Direct comparisons are possible between the kitchens to see how the range of appliances has changed.

Modern exhibitions also usually involve some form of audio-visual technology. These may take the form of a slide show accompanied by a taped commentary, video, or simply handsets which provide spoken information. The latter medium has been used very effectively in the new Plastics Gallery at the Science Museum, where taped reminiscences of workers involved in the industry bring to life the showcases of objects.

Within industrial and technological museums, demonstrations of machinery in operation have become commonplace. At GMMSI in the Electricity Gallery, small generators provide power for electric bulbs. We also operate large steam engines in our Power Hall. These include a beam engine formerly used at a colliery in Haydock. There can be little more graphic that demonstrates the grandeur of steam power than the sight of a beam engine clunking away. Unfortunately, we are not able to harness the power to drive anything. At Trencherfield Mill in Wigan, in contrast, the massive engine there still drives cotton looms that produced part of Wigan's former wealth.

Wigan Pier also uses theatre to communicate something of the flavour of nineteenth-century life. Visitors are encouraged to participate by becoming pupils in a Victorian schoolroom. A schoolmaster harangues his class according to the strict rules of behaviour demanded of children in that era. This type of living history is becoming more and more popular – at GMMSI recently schoolchildren took part in 'Railway Week', a celebration of the 1830 Liverpool–Manchester railway for which Liverpool Road Station was the original terminus.

Theatrical techniques will doubtless be used increasingly within museum exhibitions. In educational programmes also drama is being used to involve pupils/visitors in thinking through historical issues. Themes and episodes in the history of science can benefit from this kind of treatment. In the USA several museums have experimented with fairly elaborate dramatic presentations. Professional actors have been involved in developing entertainments such as 'Famous Scientist Lives Again' – 25 minutes characterizations of Charles Darwin, Isaac Newton, and Marie Curie – presented at the Pacific Science Center. In this country, the Molecule Theatre of Science for Children and productions such as Jack Klaff's *Stand Up!* are beginning to establish a tradition of problem-solving, science based drama. Neither production deals with history of science *per se*. However, there seems no reason why historical material could not be used as the basis for similar productions. The pantomime style used by the Molecule Theatre Company involves the audience in questioning and solving the practical problems which confront each character.

Museum Education Programmes

Most large museums employ trained teachers who are able to develop resource packs specifically for school groups and some of whom also offer classes. Most of these classes are gallery-based, but some may involve a short audio-visual presentation before a visit to the appropriate exhibition.

Many education services have developed practical sessions. For example, children can learn about former working methods by engaging in the technology themselves. This may involve papermaking by hand or printing using replica blocks. At GMMSI, papermaking especially is always thoroughly enjoyed, combining just the right degree of messiness with a sense of creative achievement.

Museums often organize regular lecture series. Talks may be given by curators or teachers from the institution's own staff. Invited experts also contribute frequently. There is tremendous scope for museums to cooperate with teachers. Museums are attractive locations which appeal to pupils and are generally conveniently located.

As well as trying to encourage visits, museums are also aware of the need to reach out and take their wares into the community. Curators often speak to local organizations or go into schools to give talks. Such talks may be illustrated with objects from the collection. Some museums even make objects available as loan items to members of the general public. In some instances, various objects are grouped together to form a package intended for a specific classroom use. Some institutions will lend individual items. Lending services depend upon well developed collections management and especially comprehensive cataloguing. At GMMSI, we are in the process of establishing a computer database which will enable us to identify material suitable for lending. We have already produced a guide to our electricity collection which will, we hope, help individuals determine the kind of objects they would like to borrow.

With increasing emphasis on teaching through primary sources, the use of original artefacts in the classroom may become particularly appropriate. Documents such as leaflets, company records, and photographs are generally collected by museums also. Local history libraries will also house both printed and manuscript material relating to the objects.

Museums can also be the source of visual aids, such as slides, which may be useful when developing a historical topic. Generally this kind of service is offered by education departments.

Special Events

Although the primary function of museums is to preserve our material heritage, they have also become important resources for leisure and tourism. As such, they are frequently associated with special events and exhibitions. For example in 1984 as part of the WISE (Women into Science and Engineering) initiative, we organized a series of events, including a day school for sixth form girls, a lecture by the astronomer Jocelyn Burnell, and a temporary exhibition about women scientists in history. In 1986, we hosted a day of special activities for young people including lectures and demonstrations in association with the British Association for the Advancement of Science. Museums often host steam rallies or vintage car rallies. Some of these special events may link to topics being pursued within the classroom; others may simply be useful in motivating students.

Museums are not static institutions. One of their most exciting functions is the acquisition of new objects. Sometimes, this can be quite spectacular – for example when massive objects are deposited in a store or gallery. Asking students to consider why such items become museum exhibits could be a useful way of setting historical questions.

History of science and technology are not easy subjects for pupils and students to explore without expert guidance. Museums can assist teachers in numerous ways – exhibitions can provide direct study material; curators are often willing to lecture on specialist subjects. Museums hold in trust some of the richest treasures of the nation. Their resources should be used to broaden and enliven the educational opportunities available in institutions in the United Kingdom. However, this potential will only be realized by teachers and curators working together to develop stimulating and accessible learning materials.

Major Museums of Science, Technology, and Industrial History in the UK

Science Museum
Exhibition Road
London SW7 2DD
01 589 3456

Royal Museum of Scotland
Chambers Street
Edinburgh EH1 1JF
031 225 7543

Greater Manchester Museum of
 Science and Industry
Liverpool Road Station
Castlefield
Manchester M3 4JP
061 832 2244

Birmingham Museum of Science
 and Industry
Newhall Street
Birmingham B3 1RZ
021 236 1022

National Museum of
 Photography, Film and
 Television
Prince's View
Bradford
0274 727488

Whipple Museum of the History
 of Science
Free School Lane
Cambridge CB2 3RH
0223 334540

Museum of the History of
 Science
Old Ashmolean Building
Broad Street
Oxford OX1 3AZ
0865 243997

Ironbridge Gorge Museum
Ironbridge
Telford
Shropshire TF8 7AW
095245 3522

Beamish, North of England
Open Air Museum
Beamish
nr. Stanley
Co. Durham
DH9 0RG
0207   231811

The Black Country Museum
Tipton Road
Dudley
West Midlands DY1 4SQ
021   557   9643/4

Bradford Industrial Museum
Moorside Road
Eccleshill
Bradford BD2 3HP
0274   631756

Abbeydale Industrial Hamlet
Abbeydale Road South
Sheffield S7 2QW
0742   367731

Sheffield Industrial Museum
Kelham Island
Off Alma Street
Sheffield S3 8RY
0742   22106

Leicestershire Museum of
   Technology
Abbey Pumping Station
Corporation Road
Abbey Lane
Leicestershire
0533   616330

Merseyside Maritime Museum
Pier Head
Liverpool L3 1DN
051   709   1551

Welsh Industrial and Maritime
   Museum
Bute Street
Cardiff
0222   481919

Museum of Science and
  Engineering
Blandford House
West Blandford Street
Newcastle upon Tyne
Tyne and Wear
0632   326789

Leeds Industrial Museum
Armley Mill
Canal Road
Armley
Leeds LS12 EH
0532   637861

Bristol Industrial Museum
Prince's Wharf
Prince Street
Bristol BS1 4RN
0272 299771 ext. 290

Telecom Technology Showcase
135 Queen Victoria Street
London EC4V 4AT
01 248 7444

Science Centres

Launchpad, Science Museum, London. For further information telephone
01 589 3456.

Techniquest, reopening March, 1988. For further information contact
Professor J. Beetlestone, Department of Education, University College
Cardiff. Telephone 0222 874208.

The Exploratory, Victoria Rooms, Clifton, Bristol. Telephone 0272
634321.

Xperiment! is opening in November 1988 at the Greater Manchester
Museum of Science and Industry, Liverpool Road Station, Castlefield,
Manchester M3 4JP. Telephone 061 832 2244.

Useful books

AA book of stately homes, museums, castles and gardens in Britain (1987)
Bright ideas: resource handbook (Scholastic Publications Ltd.)
British Association of Friends of Museums yearbook (annual publication)
The family guide on where to go (Jarrold 'White Horse' Series)
Sheila Grinell (ed.), *A stage for science: dramatic techniques at science-technology centers* (Association of Science and Technology Centers, Washington DC, 1979)
Donald Horne, *The great museum: the re-presentation of history* (Pluto, 1984)
K. Hudson, *The Cambridge guide to the museums of Britain and Ireland* (Cambridge University Press, 1987)
Museums yearbook (The Museums Association London)

Videos and Films

Cathy Grant

There are many hundreds of films and videos dealing with science and technology currently available for educational use from several hundred distributors in the UK. The British Universities Film and Video Council provides an information service for teachers on what programmes are available in all academic subject areas, and in 1985 we published *Discovery and Invention*, a review of 283 films and videos recommended for use in the teaching of the history of science and technology at degree level.

The following collection of 99 programmes, selected for use in schools, is to some extent complementary to that review, with 61 items being common to the two lists. The scope of the present list is narrower, in that areas such as agriculture, civil engineering and industrial archaeology have been excluded. In addition to programmes relating directly to the history of science, medicine, and technology I have included items on the philosophy of science (including scientific method), social aspects, and environmental issues, as these are increasingly finding a place in the school curriculum. In limiting the list to 99 items the main criteria used were: audience level, quality, medium (film or video), programme length, cost, and availability.

Audience level   It is often difficult to tell exactly what educational level a programme is pitched at from the information provided in the distributor's catalogue description. Many Open University programmes, for example, are suitable for sixth-form use, but I have excluded those which are obviously pitched at too high a level or appear to be too specialized. Also items which are very wide-ranging in their subject matter, e.g. the Ascent of Man series, have not been included. When programmes are aimed at or are known to be suitable for a specific age range I have indicated this in the description.

Quality   Again, it is only by reviewing a programme that its usefulness as an educational tool can really be judged. Those items marked with an asterisk (*) were reviewed by subject specialists for *Discovery and Invention* and were particularly recommended for use in introductory courses at degree level in terms of their coverage of the subject, accuracy, etc., and therefore are likely to be of use at sixth-form level. It was not possible to undertake such a reviewing project for items under consideration for inclusion in the present list, so those items without an asterisk are those which are likely to be of use with younger pupils or which have come into distribution since the *Discovery and Invention* review was undertaken.

Medium (film or video)   Preference has been given to video over film, because of its ease of use and greater flexibility. However, 16mm film may still be preferable for showing to large groups, or when attention to visual detail is particularly important. Many of the older programmes originally made on film have not been transferred to video, so there is often no choice of medium.

Programme length   It has been assumed that programmes of less than 30 minutes duration fit most readily into a teaching session, as well as being more appropriate for attention span. A few longer programmes have been included when there is little else available on that topic, or when the programme can be divided into sections. With videos, of course, it is easy to show extracts from the complete programme.

Cost   With tight budgets in mind, programmes available on free loan have been selected where possible, and those for hire included rather than those for sale only. However, matters are not quite as straightforward as this when considering the long-term pros and cons of loan or hire against purchase. If a programme is to be used repeatedly there may be advantages in buying a copy, as this avoids having to book the item several weeks or months in advance, and postage/carriage charges would not then mount up. Prices have not been included in this list since they are subject to change – a telephone call to the distributor is the best way of obtaining current accurate information. There is no such thing as the average purchase price of an educational video – a programme from ICI costs around £12.50 at the time of writing, an Open University programme £110 and a 30 minute BBC programme £99.

Availability   All the programmes listed are from well-established distributors and are available at the time of writing (February 1988). The Open University has a very interesting course entitled '**Technology and Change c.1750–1914**' (A281) which explores the technological details of industrial developments, and the social and economic context in which they occurred. Topics covered include the mechanisation of textile spinning, the use of water and steam power by the expanding industries, the making of sulphuric acid by the lead chamber process, the nineteenth century alkali industry, iron making and bridge building, the American steel industry, new industries formed as a result of new methods of burning coal, steam locomotives, the internal combustion engine, and the sugar beet processing industry. The programmes in this course have not been included in this list as it is understood that Professor Colin Russell of the Open University History Department is putting together a compilation tape of programme extracts from this course especially for use in schools. Further information on this can be obtained from Professor Russell (History Department, Open University, Walton Hall, Milton Keynes MK7 6AA. Tel. 0908 74066).

In 1984 the BBC broadcast a series called **All Our Working Lives** which drew on the memories of workers in the cotton, aircraft, steel, retail, ship-building, chemical, coal, agricultural, car, and electronics industries to provide a popular history of Britain at work since 1914. The series was accompanied by a book of the same title by Peter Pagnamenta and Richard Overy, from BBC Publications. It was hoped that these programmes would be made available for educational use, but this has not yet happened. Any enquiries about availability of this series should be addressed to BBC Television Enterprises, Educational and Training Sales, Woodlands, 80 Wood Lane, London W12 0TT. Tel. 01 743 5588 or 576 0202.

To hire or purchase any of the programmes listed below, contact the distributor directly. A list of distributors' addresses and telephone numbers is given at the end of this chapter. A licence, obtainable from Guild Sound and Vision, is necessary for those Open University items marked 'off-air recording'. Any comments or enquiries about this list should be addressed to the author: Cathy Grant, British Universities Film and Video Council, 55 Greek St, London W1V 5LR. Tel. 01 734 3687.

Programmes Arranged by Subject

The following abbreviations are used:

b&w.	black and white	p.c.	production company
col.	colour	p.i.	producing institution
dist.	distributor	sp.	sponsor
min.	minutes	st.	silent
mm.	millimetres	*	particularly recommended

Astronomy and astronautics

Cosmology before Newton*
dist. Guild Sound & Vision. Off-air recording or sale; 1980: p.c. Open University
Video, col., 22 min.
(Seventeenth century England: a changing culture 1618–1689 course)
Explains the background to Newton's *Principia*. The previously held earth-centred Aristotelian view of the universe was supplanted by a conception of the universe much as it is known now. Item no. A203/07.

Earth and Sky (2 parts)
dist. D.S. Ltd. Hire; 1958: p.i. University of Leeds
16mm film, col., 25, 20 min.
Discusses the changing conceptions of the planetary system and the heavens. Part 1 shows how early thinking in Babylonia and Greece led up to the world picture of medieval Europe. Part 2 explains how between 1500 and 1799 this conception was replaced by modern theories. Item no. 570 D6–7. *Sixth formers.*

Galileo
dist. D.S. Ltd. Hire; 1961: p.c. Coronet Films, USA
16mm film, b&w., 15 min.

Dramatized review of the life and work of Galileo. Shows how he verified the astronomical theories of Copernicus and discusses how he continued to make new discoveries in the field of physical science against opposition. Item no. 233 A14.

Kepler and his Work*
dist. D.S. Ltd. or Scottish Central Film & Video Library. Hire; 1958: p.i. Institut fuer Film und Bild, W. Germany
16mm film, b&w., 16 min.
Biographical information followed by animated diagrams showing the conception of planetary movements which was held when Kepler began his work and the revolutionary Copernican theory which Kepler supported.
Shows Kepler's work which led to the elliptic planetary movements and his three Laws of Planetary Motion. Notes provided. D.S. Item no. 572 A5; SCFVL Item no. 2UD 1901. *Mid-secondary school level.*

Kepler's War*
dist. Guild Sound & Vision. Off-air recording or sale; 1982: p.c. Open University
Video, col., 23 min.
(Inquiry course)
A study of Johannes Kepler and his Laws of Planetary Motion set against his religious and political background. Item no. U202/12.

Space Science: Exploring the Moon*
dist. D.S. Ltd. or Scottish Central Film & Video Library. Hire; 1971: p.i. Coronet Instructional Films, USA
16mm film, col., 16 min.
Chronological survey of attempts to gather knowledge of the Moon. The most important features of the Moon's surface identified before 1700. Man's knowledge of the Moon prior to the use of space probes, and the subsequent information resulting from its use. D.S. Item no. 571 D10; SCFVL Item no. 2DC 3794.

We Came in Peace*
dist. Guild Sound & Vision. Hire; 1970: p.c. CBS News, USA
16mm film, col., 38 min.
The history of spaceflight from the early experiments of Robert Goddard
to the gradual build-up for the Apollo 11 moonflight and Aldrin's and
Armstrong's lunar walk. Item no. 300 2917–2.

Biology

The DNA Story*
dist. Viscom, Hire or John Wiley, Sale; 1973
16mm film, col., 45 min.
James Watson, Francis Crick, Maurice Wilkins, and Linus Pauling
discuss the events and research leading to the discovery of the double
helix as a workable structure for DNA.

Gregor Mendel, Father of Modern Genetics
dist. German Film & Video Library. Free loan; 1963
16mm film, col., 15 min.
Covers the early stages of Mendel's research work with peas and
demonstrates his experiments to cross-breed various kinds of seed and
explain the resulting mutations. Item no. IN 0521 C(E).

Microbes and Men (6 parts)
dist. BBC Television Enterprises. Sale; 1974: p.c. BBC Television
Video, col., 55 min. each part
Part 1: Invisible Enemy Semmelweis and the search for the cause
of puerperal fever.
Part 2: Germ is Life Pasteur and Koch separately establish the link
between germs and disease.
Part 3: Men of Little Faith Pasteur's discovery of vaccines and the
hunt for the cholera germ.
Part 4: Certain Death The first successful use of vaccine against
rabies.
Part 5: Tuberculin Affair Before making proper trials Koch
announces he has found a cure for tuberculosis.
Part 6: Search for the Magic Bullet Paul Ehrlich and salvarsan –
first of the miracle drugs.

Molecules and Life*

dist. D.S. Ltd. Hire; 1969: p.c. Anvil Films for Educational Foundation for Visual Aids

16mm film, col., 20 min.

Outlines achievements in molecular biology which, over the last fifty years, have revolutionized our understanding of the chemical and physical basis of life. Shows Dorothy Hodgkin, Frederick Sanger, Max Perutz, and John Kendrew. Teaching notes available. Item no. 601 D45.

Pasteur's Experiment

dist. Educational Media International. Hire or sale; 1972

Video or 16mm film, col., 14 min.

Describes the view of spontaneous generation using a puppet scholar in classic attire performing various experiments to test its validity. Proves Pasteur's position that micro-organisms are not generated spontaneously. *Upper primary/lower secondary.*

The Pure Gamete

dist. Guild Sound & Vision. Off air-recording or sale; 1980: p.c. Open University

Video, col., 25 min.

(Science and belief: from Darwin to Einstein course)

Describes the origins of the science of genetics, concentrating on the work of Gregor Mendel and William Bateson. Discusses the problems of theory formation and development. Item no. A381/05

Botany

I, the Prince of Botany – Carolus Linnaeus*

dist. Darvill Associates. Hire; 1977: p.c. produced for Radio Sweden

16mm film, col., 20 min.

A sketch of Linnaeus' life, work, and influence which includes short interviews with such Linnaean scholars as Lindroth, Staflev, and Stearn. The commentary draws extensively on Linnaeus' own writings.

A Story of Discovery: Why Plants Bend Towards Light

dist. D.S. Ltd or Scottish Central Film & Video Library. Hire; 1966:
p.c. Encyclopaedia Britannica, USA
16mm film, col., 13 min.
Describes the contributions of Charles Darwin, Peter Boysen-Jensen,
Arpod Paal, and Frits Went in establishing the processes which cause
plants to react and bend towards the light. The isolation of hormones
has made it possible to introduce them to aid plant development.
Teaching notes available. D.S. Item no. 621 D47; SCFVL Item no.
2DC 2558

Chemistry and chemical engineering

Elements Discovered*

dist. Guild Sound & Vision. Off-air recording or sale; 1978: p.c. Open
University
Video, col., 25 min.
(Science Foundation course)
Examines how the number of known elements has changed over the
years. Includes demonstrations of classical experimental methods. Shows
the reduction of malachite by charcoal using Bronze Age type equipment
and progresses through smelting of cassiterite to tin and then to the
preparation of bronze. Shows Priestley's preparation of oxygen using
burning glass, Davy's isolation of potassium and calcium, Lockyer's
discovery of helium, and Ramsay's isolation of inert gases. Illustrates
how discovery depends on the technology available at the time. Item
no. S101/12. *A-level upwards.*

History of the Discovery of Oxygen

dist. ICI Video Library. Sale; 1949: p.c. ICI
Video, col., 16 min.
The history of combustion from the alchemists of the Phlogiston Theory,
and the final elucidation of the role of oxygen in combustion through
the work of Scheele, Priestley, and Lavoisier.

Computers

Great Experiments: Number Crunching

dist. BBC Television Enterprise. Sale; 1986: p.c. BBC Television Video,
col., 25 min.

Professor Heinz Wolff traces the history and development of the computer. Demonstrates the first mechanical counting device invented in 1642 and the principles on which it operated, and traces a succession of gradually more sophisticated machines.

Energy
Energy in Perspective
dist. BP Film Library. Free loan; 1976: p.c. Balfour Films for BP
Video or 16mm film, col., 21 min.
Discusses man's historic use of energy and examines the limits of the world's supply of fossil fuels – coal, oil, and natural gas. Considers alternative sources of energy – nuclear, solar, tidal, geothermal, etc., and emphasizes the need for conservation of energy.

Gas – Naturally
dist. BP Film Library. Free loan; 1984
Video or 16mm film, col., 22 min.
At one time, natural gas was an unwanted result of exploring for oil, so gas wells were capped. When gas was found with oil, it was burnt off on the spot. Today gas is an increasingly important source of energy. Uses a blend of light-hearted cartoon with historic and modern film to show some of the many ways in which gas is used, explains the world's reserves and rate of consumption, and points out the growing role of gas in the world energy picture.

The Great Search: Man's Need for Power and Energy
dist. Scottish Central Film & Video Library. Hire; p.c. Walt Disney, USA
16mm film, col., 12 min.
An animated film that traces man's search for energy. Shows the discovery, development, and application of the major sources of energy. Concludes by stressing man's responsibility to develop new power potentials which will not upset the ecological balance. Item no. 2WD 50.

Now and Then*
dist. British Gas Film Library. Free loan; 1976
Video or 16mm film, col., 28 min.
Describes the history of the gas industry from its introduction as an illuminating agent through to its use in heating and cooking. Discusses its competition with electricity and the factors which led to its replacement by oil and natural gas.

Offshore
dist. Shell Film Library. Free loan; 1977
Video, col., 26 min.
Traces the history of offshore oil recovery from drilling rigs mounted on barges, jack-up rigs, to semi-submersible rigs.

Thermostats
dist. British Gas Film Library. Free loan; 1980
Video or 16mm film, col., 9 min.
Describes the development of thermostats from the earlier types which were rod or linear, to the modern variety utilizing liquid bellows, concentrating on their use on domestic gas cookers. Other applications in such things as domestic refrigerators and heating systems, and various industrial processes are also considered.

Under the North Sea
dist. BP Film Library. Free loan; 1980
Video or 16mm film, col., 29 min.
A history of oil and gas from the North Sea and an examination of some of the offshore production systems of the future.

Environmental issues

Acidification – The Invisible Enemy
dist. Darvill Associates. Hire; 198-?
Video, col., 28 min.
Coal and oil burning industries, power plants, and automobile exhaust contaminate the air with chemical gases. These are borne in the wind,

and fall as acid rain, attacking vast areas of the environment in Europe and North America. There are solutions to this problem, but only if people and politicians act now.

Acid Rain – Manchester's Destructive Legacy
dist. Concord Video & Film Council. Hire or sale; 1986; p.i. Acid Rain Information Centre, Manchester Polytechnic
Video, col., 46 min.
The term 'acid rain' originated in Manchester in the 1850s. The history of damage to buildings and vegetation is discussed, and the contemporary effects of acid rain on the ecology of the region are examined. The potential dangers to human health from air pollution are highlighted. The long-distance transportation of air pollutants from Western Europe to Scandinavia is also covered.

Environment in the Balance
dist. Shell Film Library. Free loan; 1978
Video, col., 30 min.
Shows in chronological order how geological, topographical, and social developments in Britain have helped to shape our environment and discusses the problems of industrial expansion, population growth, and pollution. Creates a picture of the interaction of nature and man upon the Earth's surface. Teaching notes provided.

Industry

A Day in the Life of a Coalminer
dist. British Film Institute Film & Video Library. Hire; 1910
16mm film, st., b&w, 10 min.
A typical working day in the mines of the early 1900s. Captures the sweat and grime of the pits at a time when women and children took full share in the labour. A forerunner of today's television investigations.

Equinox: the Tin Snail
dist. Guild Sound & Vision. Sale; 1986
Video, col., 52 min.
Follows the development of the Citroen 2CV launched at the Paris Motor Show in 1948 through 40 years and 3 million cars. For the first

time since the end of the war the surviving 2CV engineers gather
together to reassess this remarkable machine.

500 years of German Printing
dist. German Film & Video Library. Free loan; 1972
16mm film, col., 8 min.
The art of book printing and development of typefaces in Germany
from the 15th century. Item no. IN 3101 B(E).

The History of Clothing (6 parts)
dist. German Film & Video Library. Free loan; 1982
Video or 16mm film, col., 28 min. each part
Six films on the history of fabrics from the earliest times to modern
industries. Titles are **Furs and Leather**, **Linen Weaving**, **Wool**,
Cotton, **Silk**, and **Synthetic Fibres**. Item nos. IN 2307–2312 B(E).

History of the Motor Car (6 parts)
dist. BP Film Library. Free loan; 1973–1979
Video or 16mm film, col., 27 min. each part
A series of 6 films tracing the development of the motor car from the
invention of the wheel until the 1970s.

In the Stream of Time
dist. German Film & Video Library. Free loan; 1984/85
Video or 16mm film, col., 15 min.
Shows how life in Germany has changed with the help of technology.
The struggle for technical perfection and also the longing to discover
new and different things to improve life. Item no. RF9436A (E).

Nine Centuries of Coal (revised edition)*
dist. British Coal. Sale; 1976
Video, col., 29 min.
The history of coal mining from the eleventh to the twentieth century.
First version made for the NCB in 1958.

The Re-birth of British Steel
dist. British Steel Information Services. Free loan; 1986: p.c. Blackrod
for British Steel Corporation
Video, col., 15 min.
Illustrates the transformation of the British Steel Corporation in less
than a decade from a loss-making operation to its present position as
efficient and profitable steelmaker. Dramatic sequences show its
contribution to industries such as nuclear energy, shipbuilding, auto-
mobiles, construction, and the oil and gas exploration and extraction
offshore industries.

The Industrial Revolution

The Industrial Revolution*
dist. D.S. Ltd. Hire; 1975
16mm film, col., 23 min.
Good illustrative material shows industrial developments that transformed the cottage industries of the eighteenth century into the power-
driven factories of the nineteenth. Includes Darby's coke iron smelting
mill, Huntsman's steel production, Bessemer's converter, the invention
and development of the steam engine. Inventions of the textile industry
are described together with developments in communications and
transport. Item no. 216 D260.

The Industrial Revolution Series (3 parts)
dist. Guild Sound & Vision. Free loan; 19—: sp. The Liberty Fund
16mm film, col., 29 min. each part
Part 1: The Great Discontinuity Concerned with the identification
and definition of the Industrial Revolution. The emphasis is on growth
as the defining characteristic, and on the entrepreneurs as the main
acts who pioneered industrialization. Item no. 346 0455–5.
Part 2: Freedom under the Law Explains why the Industrial
Revolution occurred in England and argues that the system of
government is the most important determinant of economic growth.
Item no. 344 0463–6.
Part 3: A Magnificent Century Describes the social, political, and
economic effects of the Industrial Revolution on the lives of ordinary
people. Item no. 346 0465–7.

Ironbridge and the Industrial Revolution (working title)
dist. Shell Film Library. Free loan; 1988
Video, col., 30 min. approx
At Coalbrookdale in 1709 Abraham Darby was the first man to produce
iron using coke instead of the dwindling fuel charcoal. By the end of
the eighteenth century, the area led the world in the manufacture of
iron. The working exhibits at the Ironbridge Museum are used to recreate the period, supplemented by archive material in the form of
contemporary paintings and artefacts.

Mathematics

The Birth of Calculus
dist. Guild Sound & Vision. Off-air recording or sale; 1988. p.c. Open
University
Video, col., 25 min.
(Topics in the History of Mathematics Course)
Two men independently discovered and formulated methods for the
Calculus, Newton in England and Leibniz in Paris. In this programme,
Jeremy Gray visits first Cambridge, to trace Newton's lines of thought
through his notebooks, then Hanover, where Leibniz's original notes
are stored to trace his very different approach to the same problem.
Item no. MA290/05V.

The Emergence of Greek Mathematics
dist. Guild Sound & Vision. Off-air recording or sale; 1988. p.c. Open
University
Video, col., 24 min.
(Topics in the History of Mathematics course)
Euclid's *Elements* is one of the most reprinted books of all time. How
did it come about and why does it remain a classic textbook? Starting
with the concept of 'proof', which was a relatively new one in the third
century BC, this film looks at each of the thirteen books in detail. It
shows how this classic text has been passed down to us and how
historians were able to recreate the original text from later, amended
editions. Item no. MA290/01V.

The Geometry Euclid didn't Know*
dist. Edward Patterson Associates. Hire or sale; 1979
16mm film, col., 16 min.
Traces the history of Euclid's fifth postulate, introduces elliptic and hyperbolic geometries, and makes a distinction between mathematical consistency and 'truth'.

The Idea of Numbers: an Introduction to Number Systems
dist. Edward Patterson Associates. Hire or sale; 1971
16mm film, col., 14 min.
Begins with man's earliest number concepts, such as the notion of 'one' and 'many'. Traces the development of the decimal system. With man's increasing use of numbers, the conventional arabic numeral notations have yielded to the binary system for use in computers.

The Liberation of Algebra
dist. Guild Sound & Vision. Off-air recording or sale; 1988. p.c. Open University Video, col., 25 min.
(Topics in the History of Mathematics Course)
Filmed entirely in Ireland, this programme looks at the work of William Rowan Hamilton in Dublin and George Boole in Cork. From Hamilton's discovery of quaternions to Boole's *The Laws of Thought*, Graham Flegg reviews the developments of their work in the field of astrophysics and computer design. Item MA290/08V.

Marin Mersenne – the Birth of Modern Geometry
dist. Guild Sound & Vision. Off-air recording or sale; 1988. p.c. Open University
Video, col., 25 min.
(Topics in the History of Mathematics Course)
Marin Mersenne was a French monk of the seventeenth century. He sought to break the wall of secrecy behind which scholars were hiding themselves and their discoveries. This film shows how he encouraged cooperation between scientists and how he managed to break through

the silence and publicize Descartes's development of coordinate geometry. Item MA290/03V.

Non-Euclidean Geometry

dist. Guild Sound & Vision. Off-air recording or sale; 1988. p.c. Open University
Video, col., 25 min.
(Topics in the History of Mathematics course)
During the eighteenth and nineteenth centuries mathematicians were increasingly ,questioning the foundations of geometry. This programme shows how these investigations led to the formation of non-Euclidean geometry. Item MA290/06V.

Paris and the New Mathematics

dist. Guild Sound & Vision. Off-air recording or sale; 1988. p.c. Open University.
Video, col., 24 min.
(Topics in the History of Mathematics course)
Centring on the life of one of France's greatest revolutionary mathematicians, Gaspard Monge, this programme shows how he founded the Ecole Polytechnique, originally to train military engineers. It was there that the post revolutionary mathematicians trained and did their research. Looks at the library of the present day, Ecole Polytechnique and the early nineteenth century books and papers which chronicle this period of French educational reform. Item no. MA290/07V.

The Vernacular Tradition

dist. Guild Sound & Vision. Off-air recording or sale; 1988. p.c. Open University
Video, col., 25 min.
(Topics in the History of Mathematics course)
Deals with the low-level mathematics of the Middle Ages. Compares the different notational styles of Luca Pacioli and Nicholas Chuquet. Shows how the use of the Hindu–Arabic numeral system developed and was adopted in Arabic countries and later in Europe. Traces this through the work of the Islamic mathematician Al-Kharizmi and Leonardo of Pisa (also known as Fibonacci). Item MA290/02V.

Mechanical engineering

Acting in Turn*
dist. Shell Film Library. Free loan; 1974
Video or 16mm film, col., 20 min.
Traces the development of the gear from the first recorded use as a wooden gear wheel enabling the power of harnessed animals to be used for raising water, through the watermill and windmill, mechanical clocks, steam engines, internal combustion engines, turbines, motorized transport, and the electric motor.

The History of the Oil Engine
dist. Shell Film Library. Free loan; 1961
Video or 16mm film, b&w. and col., 33 min.
Covers the period from 1861, when Otto created the atmosphere engine, to the present day. The engine was the forerunner of the first four-stroke internal combustion engine which Otto made in 1877. Shows how Herbert Ackroyd Stuart, the brothers Priestman, James Robson and others all played their part, with Diesel, in the invention of the compression–ignition engine. James McKechnie is shown with his discovery of a fuel pump, and the film describes the research of Ricardo who paved the way for modern high efficiency and lubricating oils, and the growth of the oil industry.

James Watt*
dist. D.S. Ltd. Hire; 1959: p.c. Educational Films of Scotland
16mm film, col., 16 min.
Life and work of the Scottish engineer and inventor, particularly his invention of the separate condenser. Item no. 712 A6.

The Power System
dist. Guild Sound & Vision. Hire; 1980?: sp. JCB Sales
16mm film, col., 32 min.
Uses a combination of graphic animation and location material, together with dramatizations, to provide an introduction to the development and working of hydraulic systems, and their industrial applications.

Waterwheels in Wales and the West Country*
dist. Mr J. Rogers. Hire or sale; 1978; p.c. Mr J. Rogers (Technical Films)
16mm film, col., 20 min.
Shows 16 waterwheels working in 1978 and others no longer in use. Outlines the development of waterwheels in Britain and shows some applications in cornmilling, the woollen industry, pumping, ore crushing, and driving tilt hammers and grindstones.

Werner von Siemens – Technologies of the 20th century
dist. German Film & Video Library. Free loan; 1983
Video or 16mm film, col., 26 min.
The invention of Werner von Siemens – i.e. to generate, transport, and provide electricity wherever it is needed – has revolutionized mechanical engineering. Item no. IN 2302 B(E).

Windmill
dist. Scottish Central Film & Video Library. Hire; 1908: p.c. Gateway Films
16mm film, col., 11 min.
Shows the structure and working of a well-preserved tower mill. Grain is seen being milled into wholemeal flour. Provides a clear and simple account of the working of a mill. All operations are detailed (except, curiously, the train of movement from sails through gears and shaft to stones). Item no. DC 5736.

Medicine

The Art of Physick
dist. Guild Sound & Vision. Off-air recording or sale; 1981: p.c. Open University
Video, col., 25 min.
(Seventeenth Century England: a Changing Culture 1618–1689 course)
Surveys the state of medical knowledge and treatment in the seventeenth century. Includes the compounding of herbal medicines and general discussion of chemical/Galenist physician's debates. Item no. A203/15.

Fact or Fiction: Herbal Medicine
dist. Concord Video & Films Council. Hire; 1984: p.c. Dick Young
Productions, USA
16mm film, col., 27 min.
The history of herbal medicine beginning with its documented origins
in ancient Babylonia and China, where herbal medicine has been used
for over 4000 years. Modern society's preference for the supposed
security of high-technology-produced medicine is examined in this
context.

The Fight Against Bacteria
dist. D.S. Ltd. or Scottish Central Film & Video Library. Hire; 1963
16mm film, col., 15 min.
Historical survey of methods of combating pathogenic bacteria; covers
work of Leeuwenhoek, Jenner, Pasteur, Koch, Lister, Erlich, Fleming,
and Florey and Chain at Oxford, who developed a method of producing
penicillin on a large scale. D.S. Item no. 688 D13; SCFVL Item no.
2DC 2807.

William Harvey and the Circulation of the Blood (4th edition)*
dist. Royal College of Physicians Film Library. Hire; 1978
16mm film, col., 30 min.
Animated diagrams are used to illustrate the theories of the movement
of the blood in the body up to Harvey's time. Then reconstructs Harvey's
experimental work in proof of his hypothesis put forward in *De Motu
Cordis* in 1628. The commentary is derived from his own writings.

Philosophy of science

Ethereal Fallacies
dist. Guild Sound & Vision. Off-air recording or sale; 1982; p.c. Open University
Video, col., 25 min.
(Science and Belief: from Darwin to Einstein course)
Deals with ether in nineteenth century scientific thought and in particular its significance in the work of Oliver Lodge and its connection with spiritualism. Shows how scientific discoveries can be made accidentally and how the idea of ether gains credibility through a consideration of action at a distance. Item no. A381/02.

The Founding of the Royal Society
dist. Guild Sound & Vision. Off-air recording or sale; 1988. p.c. Open University
Video, col., 24 min.
(Topics in the History of Mathematics course)
Describes the various scientific groups which began to form in seventeenth century England: at Gresham college, London in 1645 and in Oxford, 1657 culminating in the founding of the Royal Society in 1660. Looks at the problems addressed by the Society, particularly the calculation of longitude at sea which led to the foundation of the Royal Greenwich Observatory. Ends with a discussion of Newton's *Principia Mathematica*, 1687. Of interest to all those interested in the history or philosophy of science or the social history of the seventeenth century. Item no. MA290/04V.

Galileo: the Challenge of Reason*
dist. D.S. Ltd. Hire, 1970: p.c. Learning Corporation of America
16mm film, col., 26 min.
The story of Galileo, finally leading to his trial as a heretic in 1632. By dramatizing the confrontation between one individual and the Inquisition, the film explores the conflict between new scientific thinking and religious authority based on faith alone. Item no. 234 D11.

The Legacy of Wilhelm von Humboldt
dist. German Film & Video Library. Free loan; 1983

Video or 16mm film, col., 26 min.
A portrait of the Prussian who founded the first university for scientific research in 1810 in Berlin. Item no. IN 2301 B(E).

Patterns of Scientific Investigation
dist. D.S. Ltd. or Scottish Central Film & Video Library. Hire; 1964: p.c. Encyclopaedia Britannica, USA
16mm film, col., 22 min.
Attempts to provide some insight into the processes of reasoning underlying scientific discovery. Reproduces key experiments relating to static electricity to illuminate the interplay between hypothesis and experimentation, and to show the logical development of theory and method. D.S. Item no. 500 D12; SCFVL Item no. 2DC 4168.

The Philosophy of Science. Part 1*
dist. Guild Sound & Vision. Off-air recording or sale; 1982: p.c. Open University
Video, col., 24 min.
(Philosophical Problems course)
Using extracts from Microbes and Men, shows how a scientific theory is arrived at by the accumulation of a large number of factors, some of which are discarded, others used. Using the case of Semmelweis and his discovery of the cause of puerperal fever, shows that intuition and chance are important elements in scientific methodology. Item no. A313/04.

The Philosophy of Science. Part 2*
dist. Guild Sound & Vision. Off-air recording or sale; 1982: p.c. Open University
Video, col., 25 min.
(Philosophical Problems course)
Shows how scientific theory is sophisticated and modified. Begins with Semmelweis's accounting for new outbreaks of puerperal fever followed by a discussion of the theoretical insights and research of Pasteur and the laboratory methods of Koch. A useful corrective to oversimple interpretations of Popper's philosophy of science. Item no. A313/12.

Scientific Method in Action
dist. Edward Patterson Associates. Hire or sale; 1960: p.c. Visual
Education Films, USA
16mm film, col., 19 min.
Defines the scientific method, analyses its six important steps and
indicates its historical development.

Atomic physics

Atomic Physics and Reality
dist. British Universities Film & Video Council. Sale; 1985
Video, col., 40 min.
A review of the Bohr-Einstein debate on the interpretation of quantum
mechanics in the light of recent research. Bell, Aspect, Wheeler, Bohm,
and other contemporary protagonists appear and make their own
statements.

Backs to the Blast*
dist. Concord Video & Films Council. Hire; 1982: p.c. Composite
Films, Australia
16mm film, col., 50 min.
A history of the nuclear industry in South Australia which examines
the dangers of radiation. Incorporates rare archival footage and
contemporary interviews with workers, scientists, politicians, and
aborigines. Analyses the effects of uranium and its products. A uranium
mine, a uranium oxide plant and a British atomic weapons testing site
are amongst the examples shown.

Conquest of the Atom*
dist. D.S. Ltd. or Guild Sound & Vision. Hire; 1958: p.c. Realist Film
Unit for Mullard Ltd. & EFVA
Video or 16mm film, col., 22 min.
Tells the story of the structure and splitting of the atom, including the
work of J.J. Thomson, Rutherford, and Chadwick. Concludes with the
harnessing of atomic power. D.S. Item no. 557 D17; GSV Item no.
900 4451–9.

How Well we Meant
dist. Concord Video & Films Council. Hire; 1984: p.c. Immediate Family Productions, USA
Video, col., 28 min.
40 years after the invention of the atomic bomb, the scientists who had worked at Los Alamos laboratories under the direction of J. Robert Oppenheimer got together. They talk about the events that changed history, and give their thoughts about the future. Amongst those taking part are Isidor I. Rabi, Hans Bethe, Owen Chamberlain, Emilio Segre, and Edwin McMillon.

Measuring Electrons and Atoms*
dist. Guild Sound & Vision. Off-air recording or sale; 1978: p.c. Open University
Video, col., 25 min.
(Science Foundation Course)
The classic experiments to determine the nature, mass, and charge of electrons and ions are demonstrated and discussed qualitatively. Using a reconstruction of the Cavendish Laboratory, the following are shown: cathode rays, Maltese Cross experiment, Crookes paddle wheel, Millikan's oil drop, J.J. Thomson's determination from the electrostatic and magnetic deflection of an electron beam. A modern mass spectrometer is used to show the isotopes of neon. Item no. S101/10.

A New Reality*
dist. D.S. Ltd. Hire; 1962
16mm film, col., 50 min.
Traces the discovery of the structure of the atom, with particular emphasis on the work of Niels Bohr. Item no. 557 D29.

Our Friend the Atom
dist. Scottish Central Film & Video Library. Hire; 1958: p.c. Walt Disney, USA
16mm film, col., 48 min.
In two parts. Reels 1–3 trace the history of the atom from its naming by Democritus, through the invention of the telescope and the microscope. Discovery by Dalton of atomic structure of all matter, and

by Avogadro of composition of molecules; discovery of radioactivity by Becquerel, work of Curies in radium, discovery by Rutherford of chemistry of atom. Reels 4–5 cover Hahn and Strassmann probing secrets of uranium atom and accomplishing atomic fision. How atoms can be used for power, food, and health. Implications of atom in fields of transportation, medicine, agriculture, electric power. The responsibilities this source of power beings to society. Item no. 5WD 109.

Power from the Atom
dist. Viscom. Free loan; 1981: sp. United Kingdom Atomic Energy Authority
Video or 16mm film, col., 25 min.
In tracing the development of nuclear power from the discovery of the atom in the nineteenth century to the harnessing of nuclear energy for generating electricity, the film explains nuclear fission and shows how commercial nuclear reactors operate. Gas-cooled, heavy water, and pressurized water reactors are considered.

Unstable Elements
dist. Concord Video & Films Council. Hire or sale; 1985
Video, col., 90 min.
Using revealing interviews with prominent scientists and newsreels from the 1940s and 1950s, traces how Britain came to build the bomb, and how the bomb was sold to the public. As the nuclear industry continues to provoke controversy with its plans for expansion, this film reflects upon a practice which sees nature as a resource to be dominated and exploited.
NB Can be hired in two separate parts.

The World of Enrico Fermi (2 parts)
dist. Concord Video & Films Council. Hire; 1970: p.c. Visual Education Center, Canada
16mm film, col., 24 min. each part
A biography of the scientist who discovered the chain reaction in nuclear physics. Follows his career in Italy and later in America, and his meetings with other scientists including Marie Curie, Einstein,

Oppenheimer, Pauli, etc. Part 1: Italy to America. Part 2: The War Years and after.

Light

Young and the Wave Theory of Light*

dist. D.S. Ltd. Hire; 1968: p.c. Anvil Films for Department of Education and Science
16mm film, col., 21 min.
A filmed lecture given by Sir Lawrence Bragg at the Royal Institution. Outlines the two contrasting theories of light which were held in the eighteenth century. Describes and demonstrates Thomas Young's pinhole experiment. Discusses diffraction, showing its effects with the use of Young's original wave trough and explains diffraction in a series of demonstrations. Lecture notes included. Item no. 526 D40.

Magnetism

The Invisible Force*

dist. Guild Sound & Vision. Hire; 196–?: sp. Philips Industries
Video, col., 19 min.
Illustrates the ways in which scientists found that magnetism existed and how, through research, industry has revealed more of the nature of this force. Item no. 900 4459–9.

Magnetism (3 parts)

dist. D.S. Ltd. Hire; 1967: p.c. Anvil Films
16 mm film, col., 16 min. each part
(Sir Lawrence Bragg at the Royal Institution series)
Part 1 Begins with a brief survey of the history of magnetism and illustrates some of its effects. A description of the fundamental properties of magnetic materials is followed by demonstrations with the lodestone, a modern magnet, and a magnetic compass. Magnetic attraction, repulsion, and induction are next described. Discusses Faraday's contribution, reconstructing original experiments and supplementing these with demonstrations illustrating the effects of the magnetic field. Notes available. Item no. 540 D49.
Part 2 Describes Oersted's discovery of the relationship between magnetism and electricity. Proof of this is shown in a number of

demonstrations. Describes the magnetic properties of iron and illustrates its behaviour when subjected to the influence of a magentic field. Notes available. Item no. 540 D50.

Part 3 Describes and contrasts the 2 classes of magnetic bodies and demonstrates the relationship between magnetic properties and temperature. Goes on to discuss the earth's magnetism and to illustrate Wegener's theory of the history of the continents. Notes available. Item no. 540 D51.

The Development of Electrochemistry
dist. D.S. Ltd. Hire; 19 min.
16mm film, col., 19 min.
An historical survey of the discovery of electric current. Covers the work of Volta, Davy, Oersted, Faraday, and Van't Hoff. Notes available. Item no. 547 D2.

Discovering Electricity
dist. Electricity Council. Free loan; 1975
Video or 16mm film, col., 40 min.
Featuring some of Faraday's actual models and books, the film traces the story of electricity from the amber of the ancient Greeks, through lodestones and the basic discoveries of Gilbert, Franklin, Ampere, and Volta.

The Electric Wave
dist. Electricity Council. Free loan; 1982
Video or 16mm film, col., 20 min.
Begins in Faraday's laboratory in 1831 at the moment when he realized that a wave of electricity could be produced by moving a magnet in a coil of wire. A man from the future appears and takes Faraday forward into the 1980s to show him how his ideas have developed. He is shown how a giant power station produces electricity, how it is distributed, and how widely electricity is used in the modern world. Leaflet available.

Prelude to Power
dist. D.S. Ltd. Hire; 1961

16mm film, b&w., 25 min.
Deals with the life and work of Faraday, especially in the field of electromagnetism. Shows the experiments that led to the discovery of the principle of the induction ring and emphasizes the importance of this in the development of the dynamo. Notes available. Item no. 540 A28.

The Story of Magnetism
dist. D.S. Ltd. Hire; 1963: p.c. United World, USA
16mm film, col., 11 min.
Studies permanent magnets and the work of Dr Gilbert; Oersted's classical experiment showing the relationship of magnetism and electricity; Faraday's discovery of electromagnetic induction. Notes available. *Primary and secondary schools.* Item no. 540 D32.

Newton

Isaac Newton – His Life and Work
dist. Sussex Video. Sale; 1984: p.c. Imperial College
Video, col., 30 min.
Explains Newton's main scientific achievements and sets them in the context of Newton's life and times. *A-level and above.*

Newton: The Mind that Found the Future
dist. D.S. Ltd. Hire; 1971: p.c. Learning Corporation of America
16mm film, col., 21 min.
Discusses the life of Newton and his contribution to modern science based on the notion of Halley, Newton's colleague, transported into the twentieth century. Item no. 234 D12.

Relativity

Continuum*
dist. National Film Board of Canada. Sale; 1979?: p.c. NFBC
Video or 16mm film, col., 45 min.
Dr Hoffman speaks of the years he worked with Einstein at the Institute for Advanced Study in Princeton. He explains how Einstein developed his Special Theory of Relativity. Animation illustrates the basic concepts.

The concept of space-time is introduced non-mathematically, and two results of the Special Theory of Relativity are discussed – the increase of mass with increased velocity and the famous twin paradox. *A level and above.*

Einstein's Universe

dist. BBC Television Enterprises. Sale; 1979: p.c. BBC Television
Video, col., 113 min.
Documentary honouring Einstein's centenary by explaining his ideas to the layman. Among the scientists taking part are Prof. Roger Penrose, Prof. John Wheeler who knew Einstein personally, and Irwin Shapler. Also includes film from the McDonald Observatory in Texas which is involved in research into relativity.

Thermodynamics

Lord Kelvin's Clock*

dist. Guild Sound & Vision. Off-air recording or sale; 1982: p.c. Open University
Video, col., 22 min.
(Science and Belief: from Darwin to Einstein course)
Deals with Kelvin's attempts to establish scientifically the age of the earth. Shows the importance of steampower to nineteenth century science and introduces the first and second laws of thermodynamics and the idea of entropy. Item no. A381/01.

Social relations

The Great Exhibition: An Exercise in Industry

dist. Guild Sound & Vision. Off-air recording or sale; 1988. p.c. Open University
Video, col., 25 min.
(Arts Foundation course)
Paired with **The Great Exhibition: A Lesson in Taste?** A102/17v
The Great Exhibition of 1851 was intended to show the shape of things to come. But did it? This programme follows the complex and sometimes mysterious circumstances under which the Crystal Palace was designed by Joseph Paxton and shows that the construction of the palace with

its use of mass production techniques was seen as more significant than any of the innovatory exhibitions. Item no. A102/17v.

The Great Exhibition: A Lesson in Taste?
dist. Guild Sound & Vision. Off-air recording or sale; 1988. p.c. Open University
Video, col., 25 min.
(Arts Foundation course)
Paired with **The Great Exhibition: An Exercise in Industry** A102/16v
With the help of newspaper and journal reports from the time, this programme investigates how the organisers of the Great Exhibition of 1851 expected the artistic exhibits to educate the 'taste' of the visitors. In doing so, he discovers how contemporary opinions about the point and worth of the Exhibition changed while it was actually on, how important a role was played by the Royals, and how the question of what 'taste' was and how it should work, actually became a social and moral issue.

The Life and Times of Rosie the Riveter*
dist. The Other Cinema. Hire; 1981: p.c. Clarity Educational Productions, USA
Video or 16mm film, col., 65 min.
Concerned with the wartime introduction of female labour into industry to keep the shipyards, munitions factories, etc. in operation. Concentrates on five personal accounts from former female welders. Uses archive footage, contemporary songs, posters and newspaper comment to illustrate the media's attitude to women.

Making Cars
dist. Concord Video & Films Council. Hire; 1983: p.c. Television History Workshop for Channel Four Television
Video, col., 180 min.
History of the Morris Motors plant in Cowley, Oxford from 1915 to 1983, through the experiences of those who make the cars. People talk about the work they do, the skills they have learnt and lost, the struggle

to build the union, the situation of women workers, the humour and the hazards of making cars. The programme is divided into nine twenty minute segments.

A Quality of Life – Britain 1930–1960 (3 parts)
dist. Shell Film Library. Free loan; 1985
Video or 16mm film, col./b&w, 20 min. each part
GCSE, O and A level

Part 1: Industry and Technology
Shows how, in little more than a single generation, technology transformed agriculture, built the forerunners of today's computers, produced radar and the jet engine, brought about a revolution in consumer goods and services, and pioneered the peaceful uses of atomic power.

Part 2: Transport and Communication
Shows the growth of domestic and international travel during the period – an era of steam trains, airships, luxury liners and, increasingly, the aeroplane. Communication in the early '30s was via the radio, the newspaper and the cinema – later came television and, in 1957 the first satellite in space.

Part 3: Social Developments
Shows the problems of poor housing and high unemployment in the early 1930s and the first signs of improvement just prior to the Second World War, a war which brought people together in social debate, with a determination to make Britain thereafter a better place. Shows the role of the Beveridge Report in this and describes how legislation played its part in the development of the Welfare State.

Small is Beautiful
dist. Concord Video & Films Council. Hire; 1979: p.c. National Film Board of Canada
Video or 16mm film, col., 27 min.
Looks at the life and work of the economist Dr E.F. Schumacher. Shot partly in Britain where Schumacher and George McRobie of the Intermediate Technology Institute discuss and illustrate Schumacher's ideas.

Technology and the Victorian City

dist. Sussex Video. Sale; 1985
Video, col., 30 min.
The Victorians introduced municipal water, sewage, and tranport systems, and Dr Denis Smith describes how the technology involved tackled some of London's main social problems.

When the Wind Turns – Chronicle of a Work Environment

dist. German Film & Video Library. Free loan; 1980
Video or 16mm film, col., 27 min.
The Ruhr District, once a sparsely populated region, expanded rapidly following the boom in the coal industry, and for a century was a synonym for power and money. Now, it is in crisis, with almost half a million jobs lost in the mining and steel industries in the last 10 years.

Women of Steel

dist. Sheffield Film Co-op. Hire; 1984: p.c. Sheffield Film Co-op
Video, col., 27 min.
Documentary about the part played by women in the Sheffield munitions factories during the Second World War. The women talk about the jobs they did, the unions, the conditions, and how they combined all this with their home responsibilities. Interviews are intercut with archive film and stills showing the propaganda images of women munitions workers during the period.

Telecommunications

Echoes*

dist. Guild Sound & Vision. Free loan; 1982: p.c. Uden Associates for STC 16mm film, col., 25 min.
Opening in a medieval village (Monteriggione), traces the story of communication (speech, pictures, writing, printing, semaphore, telegraphy, electric telegraphy, telephone) and considers the changes which technology has made on society. Transcripts of the commentary are available. Item no. 129 1148–9.

A History of the Telephone
dist. British Telecom Film Library. Free loan; 1980
Video or 16mm film, col., 20 min.
Illustrates the changes that the telephone has undergone in the century since its invention, with a brief glimpse into the future. *Schoolchildren aged 10–14 years.*

Girdle Round the Earth
dist. D.S. Ltd. Hire; 1961: sp. Mullard and Educational Foundation for Visual Aids
16mm film, col., 20 min.
The history of the invention and development of communications by telegraph and telephone. The discovery of the electromagnet by Oersted leads on to sequences on the work of Ampere, Cook and Wheatstone, and Morse. Describes the laying of the first cross-channel cable and the two early transatlantic cables. Bell's invention of the telephone and work on wireless telegraphy with the subsequent growth of this service is shown. Finally deals with advances in cable construction and links and possibilities of using the moon and satellites for communication. Item no. 550 A7.

Transport

The Airship Hindenberg
dist. British Film Institute Film & Video Library. Hire; 1936–37
16mm film, st. b&w., 7 min.
Newsreel footage of the giant German airship in flight over New York during the 12 months in which it made 20 safe crossings of the Atlantic. Concludes with Hindenberg's abrupt collapse in flames.

The Great Highway*
dist. CFL Vision. Hire; 1965: p.c. British Transport Film Unit
Video or 16mm film, col., 21 min.
The story of the Midland main line from Stephenson to the inter-city expresses. The juxtaposition of contemporary comment and modern shots shows how little railway practice for the passenger had changed and how people viewed the new form of transport at the time. Uses contemporary pamphlets, maps, prints, and film; the air sequence and 'Coronation's' record run are of particular interest. Item no. UK 3574.

A Hundred Years Underground*

dist. CFL Vision. Hire; 1963: p.c. British Transport Films and BBC
Television
16mm film, b&w. & col., 40 min.
The development of the London underground, from the steam trains
on the Metropolitan Line in 1883 to the Victoria Line. Deals not only
with the technical development of the underground system but with its
social aspect, namely, its use as a shelter during the Second World
War. Item no. UK 3618.

Invention is not Enough

dist. Guild Sound & Vision. Off-air recording or sale; 1984: p.c. Open
University
Video, col., 24 min.
(Design: Processes and Products course)
Follows the design of the bicycle through three major phases of evolution
to illustrate the fact that a new idea has to capture public imagination,
meet a current need and find the right market in order to succeed.
Shows a selection of machines from the 1860s velocipede to a modern
recumbent design. Item no. T263/3.

The Moving Spirit

dist. BP Film Library. Free loan; 197–?
Video or 16mm film, col., 18 min.
A cartoon film tracing the development of the motor car from the
'horseless carriage' of the 1880s to the streamlined automobile of today.

Omnibus 150

dist. CFL Vision. Hire; 1979: p.c. British Transport Films for London
Transport
Video or 16mm film, col., 16 min.
Traces the origins and development of the London omnibus from
Shillebeer's three horse bus in 1829 to the familiar double-deckers of
today. Item no. UK 3611.

Sir Barnes Wallis*
dist. D.S. Ltd. Hire; 1975
16mm film, b&w, 28 min.
The life and work of Sir Barnes Wallis in the field of aviation. Begins with his early airship days and then goes on to cover the R100, the Wellington Bomber and the bouncing bomb. The second half deals with variable wing geometry and ends with a discourse on the Concorde and the F-111. Archive film and original test films are included. Item no. 237 A11.

Time Flies
dist. German Film & Video Library. Free loan; 197–?
16mm film, b&w. and col., 27 min.
The history and development of commercial airlines with particular emphasis on the German airline Lufthansa. Item no. LH 04 B(E).

The Water Highway
dist. Shell Film Library. Free loan; 1979
Video or 16mm film, col., 32 min.
The history of the canal system and the work of the Duke of Bridgewater.

Distributors' Addresses

BBC Television Enterprises
Education and Training Sales
Woodlands
80 Wood Lane
London W12 0TT
01 743 5588 or 01 576 0202

BP Film Library
15 Beaconsfield Road
London NW10 2LE
01 451 1129

British Coal
Film Library, Room 34a
Hobart House
Grosvenor Place
London SW1X 7AE
01 235 2020

British Film Institute
Film and Video Library
21 Stephen Street
London W1P 1PL
01 255 1444

British Gas Film Library
Park Hall Road Trading Estate
London SE21 8EL
01 761 3035

British Steel Information Services
9 Albert Embankment
London SE1 7SN
01 735 7654

British Telecom Film Library
25 The Burroughs
Hendon
London NW4 4AT
01 202 5342

British Universities Film &
  Video Council
55 Greek Street
London W1V 5LR
01 734 3687

CFL Vision
Chalfont Grove
Gerrards Cross
Bucks SL9 8TN
02407 4433

Concord Video and Film Council
201 Felixstowe Road
Ipswich
Suffolk IP3 9BJ
0473 715754/727747

Darvill Associates Ltd
280 Chartridge Lane
Chesham
Bucks HP5 2SG
0494 783643

D.S. Ltd
George Building
Normal College
Bangor
Gwynedd LL57 2PZ
0248 370144

Educational Media International
235 Imperial Drive
Rayners Lane
Harrow
Middlesex HA2 7HE
01 868 1908

Electricity Council Film Library
Marketing Department
30 Millbank
London SW1P 4RD
01 834 2333

Guild Sound & Vision Ltd
6 Royce Road
Peterborough PE1 5YB
0733 315 315

John Wiley & Sons Ltd
Baffins Lane
Chichester
Sussex PO19 1UD
0243 779777

Mr. J. Rogers
Technical Films
23 Rothamsted Avenue
Harpenden
Herts AL5 2DN
058 272 070

Edward Patterson Associates Ltd
Treetops
Cannongate
Hythe
Kent CT21 5PT
0303 64195

German Film & Video Library
c/o Viscom
Unit B 11
Park Hall Road Trading Estate
London SE21 8EL
01 761 4015

ICI Video Library
Thames House North
Millbank
London SW1P 4QG
01 834 4444

National Film Board of Canada
1 Grosvenor Square
London W1X 0AB
01 629 9492

Royal College of Physicians Film
    Library
c/o CFM Distribution Ltd
Pottery Lane House,
Holland Park
London W11 4LZ
01 229 5131

Scottish Central Film & Video
  Library
Dowanhill
74 Victoria Crescent Road
Glasgow G12 9JN
041 334 9314

Shell Film Library
Unit 2
Cornwall Works
Cornwall Avenue
Finchley
London N3 1LD
01 349 0025

The Other Cinema
79 Wardour Street
London W1V 3TH
01 724 8508

Sheffield Film Co-op
Albreda House
Lydgate Lane
Sheffield S10 5FH
0742 668857

Sussex Video
Townsend
Poulshot
Devizes
Wilts SN10 1SD
038 082 337

Viscom
Unit B11
Park Hall Road Trading Estate
Dulwich
London SE21 8EL
01 761 3035

References

Ahlgren, A., and J. Walberg (1973). 'Changing attitudes towards science among adolescents'. *Nature*, 245: 187–90.

Aikenhead, G.S. (1974). 'Course evaluation II: interpretation of student performance on evaluation tests'. Journal of Research in Science Teaching, II: 23–30.

Aikenhead, G.S. (1988). 'An analysis of four ways of assessing student beliefs about STS topics.' *Journal of Research in Science Teaching*, 25: 607–29.

Albury, W.R. (1977). 'From Renaissance mineral studies to historical geology, in the light of Michel Foucault's "The Order of Things"'. *British Journal for the History of Science*. 10: 187–215.

Anon. (1987). 'Why science fact is sometimes science fiction'. *The Economist*. 28 February: 103–4.

Association for Science Education (1979). *Alternatives for science education*. Hatfield.

Association for Science Education (1981). *Education through science*. Hatfield.

Association for Science Education (1986). *Science and technology in society*, units 201 (the story of Fritz Haber), 402 (DDT and malaria), 510 (Perkin's Mauve) Hatfield.

Association for Science Education (1987). *Teaching history of science and technology*. Hatfield.

Atkinson, P., and S. Delamont (1977). 'Mock-ups and cock-ups: the stage-management of guided discovery instructions.' In *School experience: explorations in the sociology of education*, ed. P. Woods and M. Hammersley. London.

Aymard, M. (1972). 'The *Annales* and French historiography'. *Journal of European Economic History*. 1: 491–511.

Bachelard, G. (1934). *Le Nouvel Esprit Scientifique*. Paris, translated as *The new scientific spirit*, Boston, Mass., 1985.

Barnes, B. (1979). *Natural order: historical studies of scientific culture*. Beverley Hills/London.

Barnes, B. (1982). *T.S. Kuhn and social science*. London.

Barnes, B., and D. Edge (1982). *Science in context: readings in the sociology of science*. Milton Keynes, Cambridge, Mass.

Becher, H.W. (1986). 'Voluntary science at Cambridge'. *British Journal for the History of Science*, 19: 57–87.

Bence-Jones, H. (1962). *Report on the past, present and future of the Royal Institution, chiefly in regard to its encouragement of scientific research*. London.

Bennett, A.J. (1964). 'Mendel's laws'. *School Science Review*, 64: 35–42.

Bennett, W.J. (1984). *To reclaim a legacy: a report on the humanities in higher education.* Washington, DC.

Bernal, J.D. (1954). *Science in history* (4 vols). London.

Berthollet, C., A. Fourcroy, A. Lavoisier, and G. de Morveau (1787). *Méthode de nomenclature chimique.* Paris.

Bevilacqua, F., and P.J. Kennedy (1983). *Proceedings of the international conference on using history of physics in innovatory physics education.* 5–9 September 1983, Pavia, Italy (Pavia: Centro studi per la Didattica della Facolta di Scienze Matematiche, Fisiche e Naturali, Università di Pavia).

Bhaskar, R. (1975). 'Feyerabend and Bachelard: two philosophies of science. *New Left Review*, 94: 31–55.

Bijker, W. (1983). 'Some Philosophical and Ethical Principle Underlying the Science Series "EXACT"'. Unpublished paper for the Workshop on Science Education and Ethics, Amsterdam.

Bloch, M. (1949). *Apologie pour l'histoire ou métier d'historien.* Paris.

Bloor, D. (1976). *Wittgenstein: a social theory of knowledge*, chap. 3. London.

Bloor, D. (1976). *Knowledge and social imagery.* London.

Bloor, D. (1978). 'Polyhedra and the abominations of Leviticus'. *British Journal for the History of Science*, 11: 245–72.

Boerhaave, H. (1735). *Elements of Chemistry*, translated by T. Dallowe. London.

Brande, W.T. (1830). A manual of chemistry, 3rd edn. London.

Brewer, J., and P. Stone (1986) *Teaching the history of science and technology – the iron industry of South Wales.* Hatfield.

British Association for the Advancement of Science (1856). *Report of the twenty-fifth meeting. . . held at Glasgow, 1855.* London.

British Association for the Advancement of Science (1918). *Report of the British Association for the Advancement of Science.* London.

Broad, W.J., and N. Wade (1982). *Betrayers of the truth.* New York.

Brock, W.H. (1985). *From protyle to proton: William Prout and the nature of matter 1785–1985.* Bristol.

Brock, W.H. (1988). 'Die Geschichte der Naturwissenschaften, an britischen Schulen: gestern, heute und morgen'. In *Zur Didaktik der Physik und Chemie: Probleme und Perspektiven*, ed. K.H. Wieble. Alsbach.

Bruner, J.S. (P1960). *The process of education.* Cambridge, Mass.

Brush, L. (1979). 'Avoidance of science and stereotypes of scientists'. *Journal of Research in Science Teaching*, 16: 237–41.

Brush, S.G. (1969). 'The role of history in the teaching of physics'. *Physics Teacher*, 7: 271–80.

Brush, S.G. (1974). 'Should the history of science be rated X?'. *Science*, 183: 1164–72.

Brush, S.G. (1979). 'Comments on "On the distortion of the history of science in science education"'. *Science Education*, 63: 277–8.

Brush, S.G. (1980). *Science Education*, 64 (1).

Brush, S.G. (1985). 'Women in physical science: from drudges to discoverers'. *Physics Teacher*, 23: 11–19.

Brush, S.G. (1987). *The history of modern science: a guide to the second scientific revolution, 1800–1950.* Ames, Ia.

Brush S.G., and A.L. King (1972). *History in the teaching of physics*. Hanover, NH.

Bud, R.F., and G.K. Roberts (1984). *Science versus practice*. Manchester.

Butterfield, H. (1931). *The Whig interpretation of history*. London.

Butterfield, H. (1944). *The Englishman and his history*. Cambridge.

Butts, D.P. (1982). 'Science education'. In *Encyclopedia of educational research*, 5th edn, ed. H.E. Mitzel, pp. 1665–75. New York.

Bynum, W.F., E.J. Browne, and R. Porter (1983). *Dictionary of the history of science*. London.

Canguilhem, G. (1963). 'The role of analogies and models in biological discovery'. In *Scientific change*, ed. A.C. Crombie, pp.507–20. London.

Canguilhem, G. (1970). *Etudes d'histoire et de philosophie des sciences*. Paris.

Canguilhem, G. (1977). *Idéologie et rationalité*. Paris.

Canguilhem, G. (1978). *On the normal and the pathological*. Dordrecht.

Canguilhem, G. (1981). 'What is scientific ideology?' Translated by M. Shortland with an 'Introduction to Georges Canguilhem'. *Radical Philosophy*, Autumn, 20–5.

Cantor, G. (1983). *Optics after Newton*. Manchester.

Cavendish, H. (1879) *The electrical researches of Henry Cavendish* (1879), ed. J.C. Maxwell, P.96. London. (Reprinted London 1967).

Chalmers, A.F. (1982). *What is this thing called science?* 2nd edn. Milton Keynes.

Champayne, A., L. Klopfer, C. Solomon, and D. Cahn (1980). *Interactions of students' knowledge with their comprehension and design of science experiments*. LRDC, University of Pittsburg.

Chant, C. (1980). *Darwin to Einstein: historical studies on science and belief*. Harlow.

Chubin, D. (1985). 'Research malpractice'. *BioScience*, 35: February, pp. 80–9.

Cohen, I.B. (1986). *The birth of a new physics*. Harmondsworth.

Collingwood, R.G. (1946). *The idea of history*. Oxford.

Collins, H.M. (1974). 'The TEA set: tacit knowledge and scientific networks', *Science Studies*, 4: 165–86.

Collins, H.M. (1975). 'The seven sexes: a study in the sociology of a phenomenon, or the replication of experiments in physics', *Sociology*, 9(2): 205–24.

Collins, H.M. (1981). 'Son of seven sexes: the social destruction of a physical phenomenon', *Social Studies of Science*, 11: 33–62.

Collins, H.M. (1982). *Sociology of scientific knowledge: a source book*. Bath.

Collins, H.M. (1985). *Changing order: replication and induction in scientific practice*. Beverly Hills/London.

Collins, H.M. (1987). 'Certainty and the public understanding of science: science on television'. *Social Studies of Science*, 17: 689–713.

Collins, H.M. and R. Harrison (1975). 'Building a TEA laser: the caprices of communication', *Social Studies of Science*, 5: 441–50.

Collins, J. and A. Smithers (1984). 'Person orientation and science choice'. *European Journal of Science Education*, 6: 55–65.

Conant, J.B. (1947). *On understanding science: an historical approach*. Oxford.

Conant, J.B. (1948). *Harvard case histories in experimental science*, 2 vols. Cambridge, Mass.

Conant, J.B. (1951). *Science and the common sense*. New Haven, Conn.

Conant, J.B. (1964). *Harvard case histories in experimental science*, vol. 1. Cambridge, Mass.

Conant, J.B. (1967). *Scientific principles and moral conduct*. Cambridge.

Concise science dictionary (1984), p.518. Oxford.

Cooter, R. (1979). 'The power of the body: the early nineteenth century'. In *Natural order: historical studies of scientific culture*, ed. B. Barnes, pp.73–92. Beverly Hills/London.

Crane, L.T. (1975). *The National Science Foundation and pre-college science education: 1950–1975*. Washington, DC.

Crawford, E. (1985). 'Learning from experience'. In *Faraday rediscovered: essays on the life and work of Michael Faraday, 1791–1867*, ed. D. Gooding and F. James, pp.211–27. London/New York.

Crombie, A.C. (1959). *Augustine to Galileo*, 2nd edn. London.

Crosland, M. (1962). *Historical studies in the language of science*. London.

Crosland, M. (1978). Gay-Lussac. Cambridge.

Crosland, M. (1980). 'Chemistry and the chemical revolution'. In *The ferment of knowledge*, ed. G. Rousseau and R. Porter. London.

Culpin, C. (1984). 'Language, learning and thinking skills in history'. *Teaching History*, 39: 24–5.

Daniell, J.F. (1839). *An introduction to the study of chemical philosophy*. London.

Dawson, I. (1987). 'School history project'. *Teaching History*, 48: 28–30.

Debus, A.G. (1978). Man and nature in the Renaissance. Cambridge.

Delaporte, F. (1982). *Nature's second kingdom: explorations of vegetality in the eighteenth century*. Cambridge, Mass.

Department of Education and Science (1985). *Science 5–16, a statement of policy*. London.

Department of Education and Science/Welsh Office (1985). *GCSE, the national criteria – chemistry*. London.

Dhombres, J. (1981). *Actes du Colloque Enseignement de l'histoire des sciences aux scientifiques*. Nantes, 1980. Paris/Nantes.

Donovan, A.L. (1975). Philosophical chemistry in the Scottish enlightenment. Edinburgh.

Donovan, A., et al. (1978). 'The history of science in undergraduate education – three approaches', *Scan: a data bank for scientific research for science faculty in schools and colleges* 2 (2): 36–41.

Douglas, M. (1966). *Purity and danger: an analysis of the concepts of pollution and taboo*. London.

Driver, R. (1983). *The pupil as scientist*. Milton Keynes.

Driver, R., Guesne, E., and Tiberghien, A. (1985). *Children's ideas in science*. Milton Keynes.

Duckworth, J. (1971). 'Imagination in teaching history'. *Teaching History*, II(5): 49–52.

Easlea, B. (1981). *Science and sexual oppression: patriarchy's confrontation with nature*. London.

Edge, D., and G. Aikenhead (1983). Paper presented to the joint seminar of the Vrije Universiteit and the World Council of Churches on Science Education and Ethics, Vrije Universiteit, Amsterdam, June.

Edwards, A.W.F. (1987). 'Defending Mendel merely perpetuating a myth'. *Nature*, 326: 449.

Ehrlich, R. (1977). 'Factors influencing secondary school enrolments'. *Physics Teacher*, 15: 490–4.

Eiseley, L.C. (1958). *Darwin's century*. New York.

Ellenius, A. (1985). 'The natural sciences and the arts'. *Acta Universitatis Upsaliensis*, 22.

Ellis, P.R. (1977). 'Alchemy to Chymistry'.

Ellis, P.R. (1981). 'The history of science and technology and the teaching of chemistry'. Unpublished M Phil thesis, University of East Anglia.

Ellis, P.R. (1987). 'Chemistry in an historical context'.

Faraday, M. (1821). 'On some new electromagnetical motions, and on the theory of magnetism.' *Quarterly journal of Science*, 12: 74–96.

Farley, J., and G. Geison (1974). 'Science, politics and spontaneous generation in nineteenth-century France: the Pasteur–Bouchet debate'. *Bulletin of the History of Medicine*, 48: 161–98.

Feyerabend, P. (1975). *Against method: outline of an anarchistic theory of knowledge*. London.

Figlio, K. (1976). 'The metaphor of organisation: an historiographic perspective on the bio-medical sciences of the early nineteenth century'. *History of Science*, 14: 17–53.

Finoocchiaro, M.A. (1980). 'A symposium on the use of history of science in the science curriculum'. *Journal of College Science Teaching*, 10 (1).

Finoochiaro, M.A., et al. (1982). 'A symposium on the use of history in the social science curriculum'. *Journal of the History of the Behavioral Sciences*, 18: 263–89.

Fleming, R. (1986). 'Adolescent reasoning in socio-scientific issues' *Journal of Research in Science Teaching*, 23(8): 677–88.

Forman, P. (1971). 'Weimar culture, causality, and quantum theory, 1918–1927: adaptation by German physicists and mathematicians to a hostile intellectual milieu'. *Historical Studies in the Physical Sciences*, 3: 1–115.

Foucault, M. (1970). *The order of things: an archaeology of the human sciences*. London.

Foucault, M. (1972). *The archaeology of knowledge*. London.

Foucault, M. (1973). *The birth of the clinic*. London.

Franklin, A. (1981). 'Millikan's published and unpublished data on oil drops'. *Historical Studies in the Physical Sciences*, 11: 185–201.

Franks, F. (1981). *Polywater*, Cambridge.

French, A.P. (1983). 'Pleasures and dangers of bringing history into physics teaching'. *Proceedings of a conference on using history of science in innovatory physics education*, 5–9 September, 1983, Pavia, Italy (Pavia: Centro Studi per la Didattica della Facolta di Scienze Matematiche, Fisiche e Naturali, Università di Pavia).

Gateshead Department of Education (1982). *King Coal*. Gateshead.

Gaukroger, S.W. (1976). 'Bachelard and the problem of epistemological analysis'. *Studies in the History and Philosophy of Science*. 7: 189–244.

Gauss, C.F. (1841). 'General propositions relating to attractive and repulsive forces acting in the inverse ratio of the square of the distances'. In *Scientific memoirs* (1841), ed. R. Taylor, Vol. iii, pp.153–96. London.

Gillispie, C.C. (1970–85). *Dictionary of scientific biography*. New York.

Gillispie, C.C. (1983). *The Montgolfier brothers*. Princeton.

Gingerich, O. (1975). '"Crisis" versus aesthetic in the Copernican revolution'. *Vistas in Astronomy*, 17: 85–93.

Gjertsen, D. (1984). *The classics of science.* New York.

Glass, B. (1959). *Forerunners of Darwin* (1745–1859). Baltimore.

Goldberg, S., et al. (1982). 'History of science in the new liberal arts curricula'. *Science technology and society.* Lehigh University.

Gooding, D. (1985a). '"In nature's school": Faraday as an experimentalist'. In *Faraday rediscovered: essays on the life and work of Michael Faraday, 1791–1867*, ed. D. Gooding and F. James. London/New York.

Gooding, D. (1985b). 'Experiment and concept-formation in electromagnetic science and technology in England, 1820–1830'. *History and Technology*, 2: 151–76.

Gooding, D. (1986). 'How do scientists reach agreement about novel observations?'. *Studies in the History and Philosophy of Science*, 17: 205–30.

Gooding, D. (1988). '"Magnetic curves" and the magnetic field: experimentation and representation in the history of a theory'. In *The uses of experiment: studies in the natural sciences*, ed. D. Gooding, T. Pinch, and S. Schaffer, pp. 182–223. Cambridge.

Gooding, D. (1988). 'Word-sketches and pictures: visual and verbal of rendering of natural phenomena'. *BSHS–HSS Joint Conference Papers*, Manchester, July, pp. 272–9.

Gooding, D. (1989). *The making of meaning: agency and observation in the language of the early field theory.* Dordrecht/Boston, Mass.

Gooding, D., and F. James (1985). *Faraday rediscovered: essays on the life and work of Michael Faraday, 1791–1867*, London/New York.

Gooding, D., T. Pinch and S. Schaffer (1988). *The uses of experiment: studies in the natural sciences.* Cambridge.

Gordon, C. (1980). 'The normal and the biological: a note on George Canguilhem'. *Ideology and Consciousness*, 7: 33–6.

Graham, L. (1981). 'Why can't history dance contemporary ballet? Or whig history and the evils of contemporary dance'. *Science, Technology and Human Values*, 6(34): 3–6.

s'Gravesande, W.J. (1747). Mathematical elements of natural philosophy. Translated by J.T. Desaguliers, 6th edn. London.

Green, G. (1828). 'An essay on the application of mathematical analyses to the theories of electricity and magnetism'. In *Mathematical papers of George Green* (1871), ed. N. Ferrers, p. 5. London.

Greene, J.C. (1959). *The death of Adam: evolution and its impact on western thought*, Iowa City, Ia.

Greene, J.C. (1981). *Science, ideology and world-view.* Berkeley, Calif.

Gruber, H. (1981). 'On the relation between "Aha experiences" and the construction of ideas'. *History of Science*, 19: 41–59

Guedon, J. C. (1977). 'Michel Foucault: the knowledge of power and the power of knowledge'. *Bulletin of the History of Medicine*, 51: 245–77.

Guerlac, H. (1977). *Essays and papers in the history of modern science*, p. 20. Baltimore.

Hacking, I. (1979). 'Michel Foucault's immature science'. *Nous*, 13: 39–51.

Haden, J.D. (1977). The use and usefulness of material of an historical nature in the teaching of elementary chemistry' Unpublished B Phil thesis, University of York.

Hall, A.R. (1948). 'Sir Isaac Newton's note-book, 1661–65'. *Cambridge Historical Journal*, IX(2): 239–50.

Hall, A.R. (1983). 'On Whiggism'. *History of Science*, 21: 45–59.

Hall, A.R. (1983). *The revolution in science, 1500–1750*, London.

Hall, A.R., and B.A. Bembridge (1986). *Physics and philanthropy – the history of the Wellcome Trust 1936–1986*. Cambridge.

Hankins, T.L. (1985). *Science and the Enlightenment*. Cambridge.

Harman, P.M. (1982). *Energy, force and matter*. Cambridge.

Harre, R. (1981). *Great scientific experiments: twenty experiments that changed our view of the world*. Oxford.

Harrison, A. (1978). *Making and thinking. A study of intelligent activities*. Hassocks/ Brighton.

Heaviside, O. (1892). *Electrical papers*. Vol. 2, p. 125. London.

Hegelson, S.L., P.E. Blosser and R.W. Howe (1977). *The status of pre-college science mathematics and social science education: 1955–1975*. Columbus, OH.

Heilbron, J. (1979). *Electricity in the seventeenth and eighteenth centuries*. Berkeley, Calif.

Heisenberg, W. (1958). *The physicist's conception of nature*. London.

Hesse, M. (1980). *Revolutions and reconstructions in the philosophy of science*. Brighton.

Hetherington, N.S. (1982). 'The history of science and the teaching of science literacy'. *Journal of Thought*, 17 (2): 53–66.

Holton, G. (1974). 'On being caught between Dionysians and Apollonians'. *Daedalus*, 103: 65–81.

Holton, G. (1978). 'Subelectrons, presuppositions and the Millikan–Ehrenhaft dispute'. *The Scientific Imagination*,. pp. 25–83. Cambridge.

Holton, G. (1978). 'On the educational philosophy of the Project Physics course'. *The scientific imagination*, pp. 284–98. New York.

Home, R.W. (1983). *Science under scrutiny: the place of history and philosophy of science*. Boston, Mass.

Home, R.W. (1983). 'Poisson's memoirs on electricity: academic politics and a new style in physics'. *British Journal for the History of Science*, 16: 239–59.

Horstmann, R.P. and M.J. Petry (1986). Hegels Philosophie der Natur. Stuttgart.

Horton, R. (1967). 'African Traditional Thought and Western Science'. *Africa*, 37: 50.

Hull, D.L. (1979). 'In defense of presentism'. *History and Theory*, 18: 1–15.

Hunt, J.A. (1979). *Revised Nuffield chemistry, chemists in the world*. Harlow.

Hunter, M. (1981). *Science and society in Restoration England*. Cambridge.

Hunter, M. (1982). *The Royal Society and its fellows*. BSHS Monograph 4.

Huxley, T.H. (1880). *The crayfish*. London.

Iggers, G.G. (1975). *New directions in European historiography*. Middletown, Conn.

Inkster, I. (1985). *The steam intellect societies*. London.

Jacob, F. (1973). *The logic of life: a history of heredity*. New York.

Jacob, J.R. (1977). *Robert Boyle and the English Revolution: a study in social and intellectual change*. New York.

Jacob, J.R. (1978). 'Boyle's atomism and the Restoration assault on pagan naturalism'. *Social Studies of Science*. 8: 211–33.

Jacob, M.C. (1976a) *The Newtonians and the English revolution 1689–1720*. Hassocks.

Jacob, M.C. (1976b) 'Seventeenth century science and religion: the state of the argument', *History of Science*, 14: 196–207.

Jennings, R.C. (1984) 'Truth, rationality and the sociology of science'. *British Journal for the Philosophy of Science.* 35: 201–11.

Jones, R. (1986). *Towards a genealogy of biology: Michel Foucault and the history of science.* Unpublished MA thesis, University of Warwick.

Jones, R.S. (1986). 'Science has become our state religion'. *Washington Post Health*, 25 June 1986.

Jordanova, L.J. (1984). *Lamarck.* Oxford.

Kauffman, G.B. (1979). 'History in the chemistry curriculum: pros and cons'. *Journal of College Science Teaching*, 36: 395–402.

King-Hele, D.G. (1985). 'The Milne Lecture'. *Quarterly Journal of the Royal Astronomical Society*, 26: 237–61.

Kirchhoff, G. (1883). *Gesammelte abhandlungen*, pp.49–55. Leipzig.

Kirchhoff, G. (1850). 'On a deduction of Ohm's laws in connexion with the theory of electrostatics'. *Philosophical Magazine*, 37: 463–8.

Klopfer, L. (1969). The teaching of science and the history of science. *Journal of Research in Science Teaching*, 6(1): 87–95.

Knight, D.M. (1984). 'The history of science in Britain; a personal view'. *Zeitschrift für Allgemeine Wissenschaftstheorie*, 15: 343–53.

Knight, D.M. (1986). 'The history of science and the history of ideas'. *History of ideas colloquium, Occasional papers* 1, pp. 22–33. Newcastle Polytechnic.

Knight, D.M. (1986). 'Accomplishment or dogma; chemistry in the introductory works of Jane Marcet and Samuel Parkes'. *Ambix*, 33: 94–8.

Knight, D.M. (1986). *The age of science*, Oxford.

Kockelmans, J.J. (1979). 'Reflections on Lakatos' methodology of scientific research programs'. In *The Structure and development of science*, ed. G. Radnitzky and G. Andersson, pp. 187–203. Boston, Mass.

Kohn, A. (1986). *False prophets: fraud and error in science and medicine.* Oxford.

Kolata, G. (1983). 'Math archive in disarray'. *Science*, 219: 940.

Koyré, A. (1939). *Etudes galiliennes.* Paris. Translated as *Galileo Studies.* Brighton.

Kramer, L.S. (1986). 'Intellectual history and reality: the search for connections'. *Historical Reflections*, 13: 517–45.

Krige, J. (1980). *Science: revolution and discontinuity.* Brighton.

Kuhn, T.S. (1962). *The structure of scientific revolutions.* Chicago/London.

Kuhn, T.S. (1963). 'The function of dogma in scientific research'. In *Scientific Change*, ed. A.C. Crombie, pp.347–69. London.

Lagrange, J.L. (1867–1892). *Oeuvres.* Vol. iv, pp.402–3. Paris.

Lamb, D., and S.M. Easton (1984). *Multiple discovery: the pattern of scientific progress.* Trowbridge.

Laudan, L. (1977). *Progress and its problems: towards a theory of scientific growth.* Berkeley/London.

Lavoisier, A.-L. (1790). *Elements of chemistry in a new systematic order. . .* Translated by R. Kerr, reprint edn, 1965, pp. 51–2. London.

Layton, D., A. Davey, and E.W. Jenkins (1986). 'Science for specific social purposes (SSSP) – perspectives on adult scientific literacy'. *Studies in Science Education*, 13: 27–52.

Lecourt, D. (1975). *Marxism and epistemology: Bachelard, Canguilhem and Foucault.* London.

Lehrman, R.L., et al. (1982). 'Physics texts: an evaluative review'. *Physics Teacher*, 20: 508–23.

Lewis, J.L. (1977). *Physics 11–13*. London.

Lewis, J.W. (1981). *Science and society*. 16 vols. London/Hatfield.

Lingenfeld, P. (1985). 'A survey of high-school physics teachers in New Jersey'. *American Journal of Physics*, 53: 1065–9.

Livingstone, R.W. (1916). *A defence of classical education*, pp. 30–1. London.

Losee, J. (1980). *A historical introduction to the philosophy of science*. 2nd edn. Oxford.

Magirus, J. (1642). *Physiologiae peripateticae libri sex cum commentariis*. Cambridge.

Mannheim, K. (1936). *Ideology and utopia*.

Mansell, A.E. (1976). 'Science for all'. *School Science Review*, 57: 579–85.

Martin, T. (1932–6). *Faraday's diary: being the various philosophical notes of experimental investigation made by Michael Faraday* . . . 7 vols + index. London.

Mas, C.J.F., J.H. Perez, and H.H. Harris (1968). 'Parallels between adolescents' conceptions of gases and the history of chemistry'. *Journal of Chemical Education*, 64(7): 616–18.

Marcuse, H. (1964). *One dimensional man*. London.

Mayr, E. (1982). *The growth of biological thought: diversity, evolution and inheritance*. Cambridge, Mass.

McEvoy, J. (1978). 'Joseph Priestley, "Aerial Philosopher": metaphysics and methodology in Priestley's chemical thought, from 1172 to 1781. *Ambix*, 25: 1–55, 93–117, 153–75; (1979) 26: 16–38.

Medawar, P.B. (1964). 'Is the scientific paper a fraud?'. In *Experiment*, ed. D. Edge, pp. 7–12, London.

Medawar, P. (1969). 'Hypothesis and imagination'. *The art of the soluble: creativity and originality in science*, p. 169. Harmondsworth.

Merton, R.K. (1938). *Science, technology, society in seventeenth century England*, published as vol.4, part 2, of *Osiris: Studies in the history and philosophy of science and on the history of learning and culture*.

Merton, R.K. (1977). 'The sociology of science: an episodic memoir'. In *The Sociology of Science in Europe*, ed. R.K. Merton and J. Gaston. Carbondale: Ill.

Millar, R., and R. Driver (1987). 'Beyond processes'. *Studies in Science Education*, 14: 33–62.

Moore, D. (1987). *Teaching the history of science*. Hatfield.

Morrell, J.B., and A. Thackray (1981). *Gentlemen of science: the early years of the British Association for the Advancement of Science*. Oxford.

Needham, J. (1954). *Science and civilisation in China*, 7 volumes to date. Cambridge.

Newton, D.P. (1986). 'Humanised science teaching and school science textbooks'. *Education Studies*, 12(1): 3–15.

Nordenskiold, E. (1928). *The history of biology*. New York.

Northern Examining Association (1988). *GCSE chemistry syllabus A for 1988 examination*.

Nuffield Foundation (1970). *Nuffield combined science*, section 2.2. London.

Ogilvie, M.B. (1986). *Women in science: antiquity through the nineteenth century*. Cambridge.

Ohm, G.S. (1841). 'The galvanic circuit investigated mathematically'. In *Scientific Memoirs* ed. R. Taylor, ii: 401–506. London.

Oldroyd, D.R. (1980). 'Sir Archibald Geikie (1835–1924), geologist, romantic aesthete, and historian of geology: the problem of whig historiography of science'. *Annals of Science*, 37: 441–62.

Oldroyd, D. (1986). *The arch of knowledge: an introductory study of the history of the philosophy and methodology of science.* London.

Pallrand, G., and P. Lingenfeld (1985). 'The physics classroom revisited: have we learned our lesson?'. *Physics Today*, 38(11): 46–52.

Partington, J. (1962). *A history of chemistry*, 4 vols. London.

Pell, A. (1985). 'Enjoyment and attainment in secondary school physics'. *British Education Research Journal*, 11(2): 123–32.

Pickstone, J. (1985). 'What is the history of science?' *History Today*, May: 46–7.

Poisson, S.D. (1812). 'Mémoire sur la distribution de l'électricité à la surface des corps conducteurs'. *Memoire de la classe des sciences mathematiques et physiques de l'Institut Impérial de France*, vol. 12, 1811, p. 14. Paris.

Porter, D.H. (1981). *The emergence of the past: a theory of historical explanation.* Chicago.

Porter, R. (1980). 'The terraqueous globe'. In *The ferment of knowledge: studies in the historiography of eighteenth century science*, ed. G.S. Rousseau. Cambridge.

Powell Davies, M. (1986). 'Using historical approaches to improve the secondary pupil's image of science'. Unpublished BA dissertation, University of Cambridge.

Priestley, J. (1782). *Disquisitions relating to matter and spirit, etc.*, 2nd edn. London.

Quattropani, D.J. (1978). 'An evaluation of the effect of Harvard Project Physics on student understanding of the relationships among science, technology, and society'. Unpublished Ph.D. dissertation, University of Connecticut, pp. 1, 62–3.

Reingold, N. (1981). 'Science, scientists and historians of science'. *History of Science.* 19: 274–8.

Ritterbush, P.C. (1964). *Overtures to biology: the speculations of eighteenth century naturalists.* New Haven, Conn.

Robinson, J.D. (1986). 'Appreciating key experiments'. *British Journal for the History of Science*, 19: 51–6.

Roche, J.J. (1987). 'Explaining electromagnetic induction: a critical re-examination'. *Physics Education*, 22: 91–9.

Roll-Hansen, N. (1983). 'The death of spontaneous generation and the birth of the gene: two case studies of relativism'. *Social Studies of Science*, 13: 481–519.

Rousseau, G.S. (1980). *The ferment of knowledge: studies in the historiography of eighteenth century science.* Cambridge.

Rudwick, M.J.S. (1985). *The great Devonian controversy.* Chicago.

Ruse, M. (1986). *Taking Darwin seriously.* Oxford.

Russell, C.A. (1982). *Science and social change 1700–1900.* London.

Russell, C.A. (1985). *Recent Developments in the history of chemistry.* London.

Russell, C.A. (1986). *Lancastrian chemist; the early years of Sir Edward Frankland.* Milton Keynes.

Russell, T.L. (1981). 'What history of science, how much, and why?'. *Science Education*, 65: 51–64.

Rutherford, F.J., G. Holton, and F.G. Watson (1970). *Project physics text.* New York.

Ryle, M. (1949). *The concept of mind.* Oxford.

Sandler, I. (1979). 'Some reflections on the protean nature of the scientific precursor'. *History of Science*, 17: 170–90.

Schaffer, S. (1980). 'Herschel in Bedlam: natural history and stellar astronomy'. *The British Journal for the History of Science*, 13: 211–39.

Schaffer, S. (1985). 'What is the history of science?' *History Today*, May: 49–50.

School Science Review (1935–6). 68: 631–2.

Scott, J. (1986). *Energy through time*. Oxford.

Secondary Science Curriculum Review (1983). *Science education 11–16. Proposals for action and consultation*. London.

Secondary Science Curriculum Review (1987). *Better science: key proposals*. London.

Shapere, D., et al. (1986). 'External and internal factors in the development of science'. *Science and Technology Studies*, 4(1): 1–23.

Shapin, S. (1979). 'The politics of observation: cerebral anatomy and social interests in the Edinburgh phrenology disputes'. In *On the margins of science: the social construction of rejected knowledge*, ed. R. Wallis, Sociological Review Monographs, 27, pp. 139–78. Keele.

Shapin, S. (1980). 'Social uses of science'. In *The ferment of knowlege: studies in the historiography of eighteenth century science*, ed. G.S. Rousseau. Cambridge.

Shapin, S. (1982). 'History of science and its sociological reconstructions'. *History of Science*, 20: 157–211.

Shapin, S. (1984). 'Pump and circumstance: Robert Boyle's literary technology'. *Social Studies of Science*, 14: 481–519.

Shapin, S., and S. Schaffer (1985). *Leviathan and the air-pump. Hobbes, Boyle and the experimental life*, pp. 1–7. Princeton.

Shemilt, D. (1980). *History 13–16 evaluation study*. Edinburgh.

Shemilt, D. (1977). *Medicine through time*, 3 vols. (revised edn 1987). Edinburgh.

Sheridan, A. (1980). *Michel Foucault, the will to truth*. London.

Sherratt, W.J. (1980). 'History of science in education'. Unpublished Ph.D. thesis, University of Leicester.

Sherratt, W.J. (1982–3). *School Science Review*, 64: 225–36, 418–24.

Shortland, M. (1981). 'Disease as a way of life'. *Ideology and Consciousness*, 9: 112–22.

Siegel, H. (1979). 'On the distortion of the history of science in science education'. *Science Education*, 63: 111–18.

Solomon, J. (1980). *Teaching children in the laboratory*, London.

Solomon, J. (1983a). 'Messy, contradictory and obstinately persistent'. *School Science Review*, 35: 225–9.

Solomon, J. (1983b). 'How can we be sure?' and 'Technology, invention and industry' *The SISCON in schools project*. Hatfield/Oxford.

Solomon, J. (1986). 'Motivation for learning science'. *School Science Review*, 67.

Solomon, J. (1987). Energy – the ghost in the body. *School Science Review*, 68: 635–44.

Solomon, J., and S. Addinell (1983). *Science in social context*. Hatfield/Oxford.

Stahl, D. (1652). *Axiomata philosophica sub titulis XX. Comprehensa*. London.

Steele, R. (1970). *Guide to the history of science.*

Stoianovich, T. (1976). *French historical method: the 'Annales' paradigm*. Ithaca, NY.

Terman, L.M. (1916). *The measurement of intelligence.*

Boston. Terman, L.M., and M.A. Merrill (1937). *Measuring intelligence*. Cambridge.

Thackray, A. (1980). 'The pre-history of an academic discipline: the study of the history of science in the United States 1891–1941'. *Minerva*, 18: 448–73.

Thomson, T. (1802). *A system of chemistry*. Edinburgh.

Thomson W. and P.G. Tait (1879). *A treatise on natural philosophy*. Cambridge.

Turner, E. (1831). *Elements of chemistry*, 3rd edn. London.

Turner, G. L'E. (1983). *Nineteenth century scientific instruments*. London.

Tyndall, J. (1884). *Faraday as a discoverer*, 4th edn. London.

Underwood, E.A. (1977). *Boerhaave's men*. Edinburgh.

University of Oxford Institute of Education Science Masters' Association (1960). *Science as a general study in the sixth form. Report of a conference of science teachers held at Oxford in July 1959*, pp. 60–4. Oxford.

Volta, A. (1779). 'Observations sur la capacité des conducteurs électriques. . .'. *Journal de Physique*, 13: 259–77.

Wallis, R.V., and P.J. Wallis (1986). *Biobibliography of British mathematics and its applications*. Newcastle upon Tyne.

Watson, W. (1747). 'A sequel to the experiments and observations tending to illustrate the nature and properties of electricity'. *Philosophical Transactions*, 44: 704–49.

Webster, C. (1975). *The great instauration. Science, medicine and reform 1626–1660*. London.

Weeks, J. (1982). 'Foucault for historians'. *History Workshop*, 14: 107–119.

Weindling, P.J. (1981). 'Periodical literature and societies'. In P. Corsi and Weindling, *Information Sources for the History of Science and Medicine*, ed. P. Corsi and P.J. Weindling, pp. 157–72, 501–8. London.

Welch, W.W. (1973). 'Review of the research and evaluation program of Harvard Project Physics'. *Journal of Research in Science Teaching*, 10: 365–78.

Welch, W.W. (1979). 'Twenty years of science curriculum development: a look back'. *Review of Research in Education*, 7: 282–306.

Welch, W., and H. Warburg (1972). 'A national experiment in curriculum evaluation'. *American Education Research Journal*, 9: 373–83.

Westfall, R. (1962). 'The foundations of Newton's philosophy of nature'. *British Journal for the History of Science*, 1(2): 171–82.

Whitaker, M.A.B. (1984). 'Science, scientists and history of science. *History of Science*, 22: 421–4.

Whittaker, E. (1961–2). *A history of the theories of aether and electricity*, 2 vols. London.

Williams, T.I. (1969). *A biographical dictionary of scientists*. London.

Wintle, J. (1982). *Makers of nineteenth-century culture*. London.

Young, R.M. (1969). 'Malthus and the evolutionists: the common context of biological and social theory'. *Past and Present*, 43: 109–41.

Young, R.M. (1973). 'The historiographic and ideological context of the nineteenth century debate on man's place in nature'. In *Changing perspectives in the history of science*, ed. M. Teich and R. Young, pp. 344–438. London.

Young, R.M. (1979). 'Interpreting the production of science'. *New Scientist*, 29 March.

Yoxen, E.J. (1985). 'Speaking out about competition: an essay on *The double helix* as popularisation. In *Expository science: forms and functions of popularisation*, ed. T. Shinn, R.D. Whitley, pp. 162–81. Dordrecht.

Ziman, J. (1976). *The force of knowledge*. Cambridge.

Ziman, J. (1980). *Teaching and learning about science and society*, pp. 118–23. Cambridge.